Valve Selection Handbook

Second Edition

Gulf Publishing Company
Book Division
Houston, London, Paris, Tokyo

Valve Selection Handbook

Second Edition

R. W. Zappe

Valve Selection Handbook

Second Edition

Library of Congress Cataloging-in-Publication Data

Zappe, R. W., 1912–
  Valve selection handbook.

  Bibliography: p.
  Includes index.
  1. Valves—Handbooks, manuals, etc. I. Title.
TS277.Z36 1987   621.8′4   86-25646

ISBN 0-87201-918-7

The information, opinions and recommendations in this book are based on the author's experience and review of the most current knowledge and technology, and are offered solely as guidance on the selection of valves for the process industries. While every care has been taken in compiling and publishing this work, neither the author nor the publisher can accept any liability for the actions of those who apply the information herein.

Dedication

to
William Hanke
brother-in-law and fellow engineer
in memory

Contents

Preface

Valves are the controlling elements in fluid flow and pressure systems. Like many other engineering components, they have developed over some three centuries from primitive arrangements into a wide range of engineered units satisfying a great variety of industrial needs.

The wide range of valve types available is gratifying to the user, for the chance is very high that a valve exists which matches his application. But because of the apparently innumerable alternatives, the user must have the knowledge and skill to analyze each application and determine the factors on which the valve can be selected. He must also have sufficient knowledge of valve types and their construction to make the best selection from those available.

Reference manuals on valves and their construction and use are readily available. But there exist few, if any, books devoted to valves which discuss the engineering fundamentals related to them, or which deal in depth with the factors on which the selection should be made.

This volume is the result of a lifetime of study in the field of design and application of valves, and it guides the user on the selection of valves by analyzing valve duty and valve construction. The book is meant to be a reference text for practicing engineers and students, but it will also be of interest to manufacturers of valves, statutory authorities, and others.

The valves discussed in this book are manual valves, check valves, and pressure relief valves. There is also an important section on rupture discs. An outline of the scope and approach of this book may be found in Chapter 1, "Introduction."

It has been found expedient to revise some chapters and introduce new ones. This second edition also provides a welcome opportunity to correct a few minor errors that escaped the most careful proofreading and seem to be inherent in first editions.

Following are some of the revised sections: Flow through Valves, Wedge Gate Valves, Ball Valves, Butterfly Valves, and Check Valves. The revision of Flow

Through Valves follows the presentation in the ISA Standards S.39.1 through S.39.4 and the publications 534-1 and 534-2 of the International Electrochemical Commission.

Some revisions have also been made to the presentation of the sizing equations for pressure relief valves. A sizing equation for liquids flashing from the fully liquid state when passing through the valve has been added.

Of particular interest is a case study in the section on wedge gate valves which deals with a massive failure of wedge gate valves in a large petrochemical plant.

New topics include gaskets of exfoliated graphite and rupture discs. Rupture discs are non-reclosing pressure relief devices and, therefore, cannot be classified as valves. However, rupture discs fulfill the same pressure relieving function as pressure relief valves. They are frequently used in parallel or in series with pressure relief valves. Because of the relationship between pressure relief valves and rupture discs, this book now includes a thorough discussion of rupture discs.

I would like to thank again the many persons and companies who so freely offered their advice and gave permission to use their material.

R.W. Zappe

1

Introduction

Valves are the components in a fluid flow or pressure system which regulate either the flow or the pressure of the fluid. This duty may involve stopping and starting flow, controlling flow rate, diverting flow, preventing back flow, controlling pressure, or relieving pressure.

These duties are performed by adjusting the position of the closure member in the valve. This may be done either manually or automatically; manual operation includes also operation of the valve by means of a manually controlled power operator. The valves discussed here are manually operated valves for stopping and starting flow, controlling flow rate, and diverting flow; and automatically operated valves for preventing back flow and relieving pressure. The manually operated valves are referred to as manual valves, while valves for the prevention of back flow and the relief of pressure are referred to as check valves and pressure relief valves, respectively.

Rupture discs are non-reclosing pressure relieving devices which fulfill a duty similar to pressure relief valves.

Fundamentals

The sealing performance and the flow characteristics are important aspects in valve selection. An understanding of these aspects is helpful and often essential in the selection of the correct valve. Chapter 2 deals with the fundamentals of valve seals and flow through valves.

The discussion on valve seals begins with the definition of fluid tightness, followed by a description of the sealing mechanism and the design of seat seals, gasketed seals, and stem seals. The subject of flow through valves covers pressure loss, cavitation, waterhammer, and attenuation of valve noise.

1

Manual Valves

The manual valves are divided into four groups according to the way the closure member moves onto the seat. Each valve group consists of a number of distinct types of valves which, in turn, are made in numerous variations.

The way the closure member moves onto the seat gives a particular group or type of valve a typical flow-control characteristic. This flow-control characteristic has been used to establish a preliminary chart for the selection of valves. The final valve selection may be made from the description of the various types of valves and their variations which follows that chart.

Check Valves

The many types of check valves are likewise divided into four groups according to the way the closure member moves onto the seat.

The basic duty of these valves is to prevent back flow. However, the valves should also close fast enough to prevent the formation of a significant reverse-flow velocity which, on sudden shut-off, may introduce an undesirably high surge pressure and/or cause heavy slamming of the closure member against the seat. In addition, the closure member should remain stable in the open valve position.

Chapter 4, on check valves, describes the design and operating characteristics of these valves, and discusses the criteria upon which check valves should be selected.

Pressure Relief Valves

There are two major groups of pressure relief valves: direct-acting pressure relief valves which are actuated directly by the pressure at the valve inlet, and piloted pressure relief valves in which a pilot mechanism opens and closes a main valve in response to the pressure in the pressure system.

Most pressure relief valves commonly used are the direct-acting type. Their development and design from primitive to modern types is discussed. Particular attention is given to explain the influence of inlet pressure loss and back pressure on the performance of pressure relief valves. This is done through performance diagrams which display the lifting and closing forces against the valve lift.

The piloted pressure relief valves have become important for large steam power plants, and to some extent for refineries and for chemicals and related industries. The valves may be operated by the system fluid or an external medium. The various design and operating principles which permit fail-safe operation of these valves are discussed, followed by installation diagrams accepted by certain statutory authorities.

Methods are also given for finding the size of pressure relief valves and for calculating inlet pressure loss, back pressure, discharge reactive force, and the size of exhaust silencer. These methods are illustrated by sample calculations, and some equations are derived.

Rupture Discs

Rupture discs are non-reclosing pressure relief devices, which may be used alone or in conjunction with pressure relief valves. The principal types of rupture discs are tension and compression rupture discs; one fails in tension and the other in compression.

Tension-type rupture discs require a fairly high margin between operating and burst pressure, in contrast with compression-type rupture discs, which can be operated at pressures close to the burst pressure without fatigue failure. However, not all compression-type rupture discs are suitable for totally full liquid systems.

Rupture discs may also be provided with burst indicators, which, in addition to monitoring rupture disc failure, may be used to activate or deactivate pumps, valves, or systems in response to the failure of rupture discs.

Units of Measurement

Measurements are given in SI and imperial units. Numerical values in equations are collected in a constant, B, which applies to the units of measurement used in solving the equation.

Identification of Valve Size and Pressure Class

The identification of valve sizes and pressure classes in this book follows the recommendations contained in MSS Standard Practice SP-86.

Nominal valve sizes and pressure classes are expressed without the addition of units of measure; e.g., NPS 2, DN 50 and Class 150, PN 20. NPS 2 stands for nominal pipe size 2 in. and DN 50 for diameter nominal 50 mm. Class 150 stands for class 150 lb and PN 20 for pressure nominal 20 bar.

Standards

Appendix C contains the more important U.S., British, and ISO standards pertaining to valves. The standards are grouped according to valve type or group.

2

Fundamentals

Fluid Tightness of Valves

Valve Seals

One of the duties of most valves is to provide a fluid seal between the seat and the closure member. If the closure member is moved by a stem that penetrates from the outside into the pressure system, another fluid seal must be provided around the stem. Seals must also be provided between the pressure-retaining valve components. If the escape of fluid into the atmosphere cannot be tolerated, the latter seals can assume a higher importance than the seat seal. Thus, the construction of the valve seals can greatly influence the selection of valves.

Leakage Criterion

A seal is fluid tight if the leakage is not noticed or if the amount of noticed leakage is permissible. The maximum permissible leakage for the application is known as the leakage criterion.

The fluid tightness may be expressed either as the time taken for a given mass or volume of fluid to pass through the leakage capillaries, or as the time taken for a given pressure change in the fluid system. Fluid tightness is usually expressed in terms of its reciprocal, i.e. leakage rate or pressure change.

Four broad classes of fluid tightness for valves can be distinguished: nominal-leakage class, low-leakage class, steam class, and atom class.

The nominal- and low-leakage classes apply only to the seat seal of valves which are not required to shut off tightly, as in the case of some for the control of flow rate. Steam-class fluid tightness is relevant to the seat, stem, and body-joint seals of valves used for steam and most other industrial applications. Atom-class fluid tightness applies to situations in which an extremely high degree of fluid tightness is required, as in spacecraft and atomic power plant installations.

Lok[1] introduced the terms steam class and atom class for the fluid tightness of gasketed seals, and proposed the following leakage criteria.

Steam class:
    Gas leakage rate 10 to 100 μg/s per meter seal length.
    Liquid leakage rate 0.1 to 1.0 mg/s per meter seal length.

Atom class:
    Gas leakage rate 10^{-3} to 10^{-5} μg/s per meter seal length.

In the USA, atom-class leakage is commonly referred to as zero leakage. A definition of zero leakage for spacecraft requirements is contained in a technical report of the Jet Propulsion Laboratory, California Institute of Technology.[26] Accordingly, zero liquid leakage exists if surface tension prevents the entry of liquid into leakage capillaries. Zero gas leakage as such does not exist. An arbitrary curve shown in Figure 2-1 was constructed for use as a specification standard for zero gas leakage.

Proving Fluid Tightness

Most valves are intended for duties for which steam-class fluid tightness is satisfactory. Tests for proving this degree of fluid tightness are normally carried out with water, air, or an inert gas. The tests are applied to the valve body and the seat, and, depending on the construction of the valve, also to the stuffing-box back seat, but they exclude frequently the stuffing-box seal itself.

When testing with water, the leakage rate is metered in terms of either volume per time unit or liquid droplets per time unit. Gas leakage may be metered by conducting the leakage gas through either water or a bubble-forming liquid leak-detector agent, and then counting the leakage gas bubbles per time unit. Using a bubble-forming liquid leak-detector agent permits metering very low leakage rates, down to 1×10^{-2} or 1×10^{-4} sccs (standard cubic centimeters per second), depending on the skill of the operator.[35]

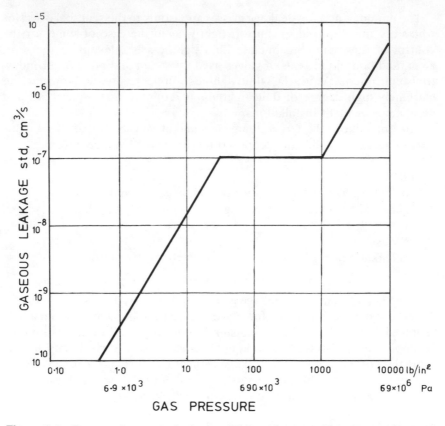

Figure 2-1. Proposed zero gas leakage criterion. (Courtesy of Jet Propulsion Laboratory, California Institute of Technology. Reproduced from JPL Technical Report No. 32-926.)

Lower leakage rates in the atom class may be detected by using a search gas in conjunction with a search-gas detector.

Specifications for proving leakage tightness may be found in valve standards or in the separate standards listed in Appendix C. A description of leakage testing methods for the atom class may be found in BS 3636.

Sealing Mechanism

Sealability Against Liquids

The sealability against liquids is determined by the surface tension and the viscosity of the liquid.

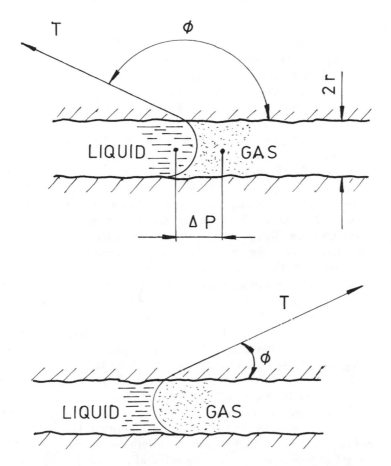

Figure 2-2. Effect of surface tension on leakage flow through capillary.

When the leakage capillary is filled with gas, surface tension can either draw the liquid into the capillary or repel the liquid, depending on the angle of contact formed by the liquid with the capillary wall. The value of the contact angle is thereby a measure of the degree of wetting of the solid by the liquid, and it indicates the relative strength of the attractive forces exerted by the capillary wall on the liquid molecules, compared with the attractive forces between the liquid molecules themselves.

Figure 2-2 illustrates the forces acting on the liquid in the capillary. The opposing forces are in equilibrium if

$$\pi r^2 \, \Delta P \; = \; 2\pi \, rT\cos\theta \quad \text{or} \quad \Delta P \; = \frac{2T\cos\theta}{r} \qquad (2\text{-}1)$$

where

r	= radius of capillary
ΔP	= capillary pressure
T	= surface tension
θ	= contact angle between the solid and liquid

Thus, if the contact angle formed between the solid and liquid is greater than 90°, surface tension can prevent leakage flow. Conversely, if the contact angle is less than 90°, the liquid will draw into the capillaries and leakage flow will start at low fluid pressures.

The tendency of metal surfaces to form a contact angle with the liquid of greater than 90° depends on the presence of a layer of oily, greasy, or waxy substances which normally cover metal surfaces. When this layer is removed by a solvent, the surface properties alter, and a liquid which previously was repelled may now wet the surface. For example, kerosene dissolves a greasy surface film, and a valve which originally was fluid tight against water may leak badly after the seatings have been washed with kerosene. Wiping the seating surfaces with an ordinary cloth can be sufficient to restore the greasy film and, thus, the original seat tightness of the valve against water.

Once the leakage capillaries are flooded, the capillary pressure becomes zero unless gas bubbles carried by the fluid break the liquid column. If the diameter of the leakage capillary is large, and the Reynolds number of the leakage flow higher than critical, the leakage flow is turbulent. As the diameter of the capillary decreases and the Reynolds number decreases below its critical value, the leakage flow becomes laminar. This leakage flow will, from Poisuille's equation, vary inversely with the viscosity of the liquid and the length of the capillary, and proportionally to the driving force and the diameter of the capillary. Thus, for conditions of high viscosity and small capillary size, the leakage flow can become so small that it reaches undetectable amounts.

Sealability Against Gases

The sealability against gases is determined by the viscosity of the gas and the size of the gas molecules. If the leakage capillary is large, the leakage flow will be turbulent. As the diameter of the capillary decreases and the Reynolds number decreases below its critical value, the leakage flow becomes laminar, and the leakage flow will, from Poisuille's equation, vary inversely with the viscosity of the gas and the length of the capillary, and proportionally to the driving force and the diameter of the capillary. As the diameter of the capillary decreases still further until it is of the same order of magnitude as the free mean path of the gas molecules, the flow loses its mass character and becomes diffusive, i.e. the gas molecules flow through the capillaries by random thermal motion. The size of the capillary may decrease finally below the molecular size of the gas, but even then, flow will not strictly cease, since gases are known to be capable of diffusing through solid metal walls.

Mechanism for Closing Leakage Passages

Machined surfaces have two components making up their texture: a waviness with a comparatively wide distance between peaks, and a roughness consisting of very small irregularities superimposed on the wavy pattern. Even for the finest surface finish, these irregularities are large compared with the size of a molecule.

If the material of one of the mating bodies has a high enough yield strain, the leakage passages formed by the surface irregularities can be closed by elastic deformation alone. Rubber, which has a yield strain of approximately 1,000 times that of mild steel, provides a fluid-tight seal without being stressed above its elastic limit. Most materials, however, have a considerably lower elastic strain, so the material must be stressed above its elastic limit to close the leakage passages.

If both surfaces are metallic, only the summits of the surface irregularities meet initially, and small loads are sufficient to deform the summits plastically. As the area of real contact grows, the deformation of the surface irregularities becomes plastic-elastic. When the gaps formed by the surface waviness are closed, only the surface roughness in the valleys remains. To close these remaining channels, very high loads which can cause severe plastic deformation of the underlying material must be applied. However, the intimate contact between the two faces needs to extend only along a continuous line or ribbon to produce a fluid-tight seal. Radially directed asperities are difficult or impossible to seal.

Valve Seatings

Valve seatings are the portions of the seat and closure member which contact each other for closure. Because the seatings are subject to wear during the making of the seal, the sealability of the seatings tends to diminish with operation.

Metal Seatings

Metal seatings are prone to deformation by trapped fluid and wear particles. The seatings are further damaged by corrosion, erosion, and abrasion. If the wear-particle size is large compared with the size of the surface irregularities, the surface finish will deteriorate as the seatings wear in. On the other hand, if the wear-particle size is small compared with the size of the surface irregularities, a coarse finish tends to improve as the seatings wear in. The wear-particle size depends thereby not only on the type of the material and its condition, but also on the lubricity of the fluid and the contamination of the seatings with corrosion and fluid products, both of which reduce the wear-particle size.

The seating material must therefore be selected for resistance to erosion, corrosion, and abrasion. If the material fails in one of these requirements, it may be completely unsuitable for its duty. For example, corrosive action of the fluid greatly accelerates erosion. Similarly, a material which is highly resistant to erosion and corrosion may fail completely because of poor galling resistance. On the other hand, the best material may be too expensive for the class of valve being considered, and a compromise may have to be made.

Table 2-1 gives data on the resistance of a variety of seating materials to erosion by jets of steam. Stainless steel AISI type 410 (13 Cr) in heat-treated form is shown to be particularly impervious to attack from steam erosion. However, if the fluid lacks lubricity, type 410 stainless steel in like contact offers only fair resistance to galling. The galling tendency, though, is reduced if the mating components are of different hardness. For steam and other fluids which lack lubricity, a combination of type 410 stainless steel and copper-nickel alloy is frequently used. Stellite, a cobalt-nickel-chromium alloy, has proved most successful against erosion and galling at elevated temperatures, and against corrosion for a wide range of corrosives.

A list of seating materials and their combinations frequently used in steel valves may be found in API Std 600.

Sealing with Sealants

The leakage passages between metal seatings can be closed by sealants injected into the space between the seatings after the valve has been closed. One metal-seated valve which relies completely on this sealing method is the lubricated plug valve. The injection of a sealant to the seatings is used also in some other types of valves to provide an emergency seat seal after the original seat seal has failed.

Soft Seatings

In the case of soft seatings, one or both seating faces may consist of a soft material such as plastic or rubber. Because these materials conform readily to the mating face, soft-seated valves can achieve an extremely high degree of fluid tightness. Also, the high degree of fluid tightness can be achieved

Table 2-1
Erosion Penetration
(Courtesy Crane Co.)

Resulting from the impingement of a 1.59 mm (1/16 inch) diameter jet of saturated steam of 2.41 MPa (350 psi) pressure for 100 hours on to a specimen 0.13 mm (0.005 inch) away from the orifice:

Class 1 — less than 0.0127 mm (0.0005 inch) penetration

Stainless steel AISI tp 410 (13Cr) bar forged and heat treated
Delhi hard (17Cr)
Stainless steel AISI tp 304 (18Cr, 10Ni) cast
Stellite No. 6

Class 2 — 0.0127 mm (0.0005 inch) to 0.0254 mm (0.001 inch) penetration

Stainless steel AISI tp 304 (18Cr, 10Ni) wrought
Stainless steel AISI tp 316 (18Cr, 12Ni, 2.5Mo) arc deposit
Stellite No.6 torch deposit

Class 3 — 0.0254 mm (0.001 inch) to 0.0508 mm (0.002 inch) penetration

Stainless steel AISI tp 410 (13Cr) forged, hardened 444 Bhn
Nickel—base copper—tin alloy
Chromium plate on No.4 brass (0.0254 mm = 0.001 inch)

Class 4 — 0.0508 mm (0.002 inch) to 0.1016 mm (0.004 inch) penetration

Brass stem stock
Nitralloy 2½ Ni
Nitralloy high carbon and chrome
Nitralloy Cr—V sorbite—ferrite lake structure, annealed after nitriding 950 Bhn
Nitralloy Cr—V Bhn 770 sorbitic structure
Nitralloy Cr—Al Bhn 758 ferritic structure
Monel modifications

Class 5 — 0.1016 mm (0.004 inch) to 0.2032 mm (0.008 inch) penetration

Brass No.4, No.5, No.22, No.24
Nitralloy Cr—Al Bhn 1155 sorbitic structure
Nitralloy Cr—V Bhn 739 ferrite lake structure
Monel metal, cast

Class 6 — 0.2032 mm (0.008 inch) to 0.4064 mm (0.016 inch) penetration

Low alloy steel C 0.16, Mo 0.27, Si 0.19, Mn 0.96
Low alloy steel Cu 0.64, Si 1.37, Mn 1.42
Ferro steel

Class 7 — 0.4064 mm (0.016 inch) to 0.8128 mm (0.032 inch) penetration

Rolled red brass
Grey cast iron
Malleable iron
Carbon steel 0.40 C

Table 2-2
Experimentally Determined Temperature Rise for Oxygen Due to Sudden Pressurizing from an Initial State of Atmospheric Pressure and 15°C.

Sudden Pressure Rise		Temperature Rise	
2.5	MPa (360 lb/in²)	375°C	(705°F)
5	MPa (725 lb/in²)	490°C	(915°F)
10	MPa (1450 lb/in²)	630°C	(1165°F)
15	MPa (2175 lb/in²)	730°C	(1345°F)
20	MPa (2900 lb/in²)	790°C	(1455°F)

repeatedly. On the debit side, the application of these materials is limited by their degree of compatibility with the fluid and by temperature.

A sometimes unexpected limitation of soft seating materials exists in situations in which the valve shuts off a system that is suddenly filled with gas at high pressure. The high-pressure gas entering the closed system acts like a piston on the gas which filled the system. The heat of compression can be high enough to disintegrate the soft seating material.

Table 2-2 indicates the magnitude of the temperature rise which can occur. This particular list gives the experimentally determined temperature rise for oxygen which has been suddenly pressurized from an initial state of atmospheric pressure and 15°C.[17]

Heat damage to the soft seating element is combatted in globe valves by a heat sink resembling a metallic button with a large heat-absorbing surface, which is located ahead of the soft seating element. In the case of oxygen service, this design measure may not be enough to prevent the soft seating element from bursting into flames. To prevent such valve failure, the valve inlet passage may have to be extended beyond the seat passage, so that the end of the inlet passage forms a pocket in which the high temperature gas can accumulate away from the seatings.

In designing soft seatings, the main consideration is to prevent the soft seating element from being displaced or extruded by the fluid pressure.

Gaskets

Flat Metallic Gaskets

Flat metallic gaskets adapt to the irregularities of the flange face by elastic and plastic deformation. To inhibit plastic deformation of the flange face, the yield shear strength of the gasket material must be considerably lower than that of the flange material.

The free lateral expansion of the gasket due to yielding is resisted by the roughness of the flange face. This resistance to lateral expansion causes the yield zone to enter the gasket from its lateral boundaries, while the remainder of the gasket deforms initially elastically. If the flange face is rough enough to prevent slippage of the gasket altogether — in which case the friction factor is 0.5 — the gasket will not expand until the yield zones have met in the center of the gasket.[5]

For gaskets of a non-strain-hardening material mounted between perfectly rough flange faces, the mean gasket pressure is, according to Lok[1], approximately:

$$P_m = 2k \left(1 + \frac{w}{4t}\right) \qquad (2\text{-}2)$$

where
P_m = mean gasket pressure
k = yield shear stress of gasket material
w = gasket width
t = gasket thickness

If the friction factor were zero, the gasket pressure could not exceed twice the yield shear stress. Thus, a high friction factor improves the load-bearing capacity of the gasket.

Lok has also shown that a friction factor lower than 0.5 but not less than 0.2 diminishes the load-bearing capacity of the gasket only by a small amount. Fortunately, the friction factor of finely machined flange faces is higher than 0.2. But the friction factor for normal aluminum gaskets in contact with lapped flange faces has been found to be only 0.05. The degree to which surface irregularities are filled in this case is very low. Polishing the flange face, as is sometimes done for important joints, is therefore not recommended.

Lok considers spiral grooves with an apex angle of 90° and a depth of 0.1mm (250 grooves per inch) representative for flange face finishes in the steam class, and a depth of 0.01mm (2500 grooves per inch) representative in the atom class. To achieve the desired degree of filling of these grooves, Lok proposes the following dimensional and pressure-stress relationships.

for steam class: $\quad \dfrac{w}{t} > 5 \quad$ and $\quad \dfrac{P_m}{2k} > 2.25$

for atom class: $\quad \dfrac{w}{t} > 16 \quad$ and $\quad \dfrac{P_m}{2k} > 5$

Compressed Asbestos Fiber Gaskets

Compressed asbestos fiber is designed to combine the properties of rubber and asbestos. Rubber has the ability to follow readily the surface irregularities of the flange face, but it cannot support high loads in plain strain or withstand higher temperatures. To increase the load-carrying capacity and temperature resistance of rubber, but still retain some of its original property to accommodate itself to the mating face, the rubber is reinforced with asbestos fiber. To these materials, binders, fillers, and colors are added.

This composition contains fine capillaries which are large enough to permit the passage of gas. The numbers and sizes of the capillaries vary for product grades, and tend to increase with decreasing rubber content. Reinforcing wire which is sometimes provided in compressed asbestos fiber gaskets tends to increase the permeability of the gasket to gas. Consequently, an optimum seal against gas will result when not only the irregularities of the flange faces are closed but also the capillaries in the gasket. To close these capillaries, the gasket must be highly stressed.

The diffusion losses can be combatted by making the compressed asbestos gasket as thin as possible. The minimum thickness depends on the surface finish of the flange face and the working stress required for the gasket to conform to the surface irregularities while still retaining sufficient resiliency. Because the properties of compressed asbestos vary between makes and product grades, the manufacturer must be consulted for design data such as provided by the publication in Reference 2.

Gaskets of Exfoliated Graphite[77]

Exfoliated graphite is manufactured by the thermal exfoliation of graphite intercalation compounds and then calendered into flexible foil and laminated without an additional binder. The material thus produced possesses extraordinary physical and chemical properties that render it particularly suitable for gaskets. Some of these properties are:

- High impermeability to gases and liquids, irrespective of temperature and time.
- Resistance to extremes of temperature, ranging from $-200\,°C$ ($-330\,°F$) to $500\,°C$ ($930\,°F$) in oxidizing atmosphere and up to $3000\,°C$ ($5430\,°F$) in reducing or inert atmosphere.
- High resistance to most reagents, for example, inorganic or organic acids and bases, solvents, and hot oils and waxes. (Exceptions are strongly oxidizing compounds such as concentrated nitric acid, highly concentrated sulfuric acid, chromium(VI)-permanganate solutions, chloric acid, and molten alkaline and alkaline earth metals.)

- Graphite gaskets with an initial density of 1.0 will conform readily to irregularities of flange faces, even at relatively low surface pressures. As the gasket is compressed further during assembly, the resilience increases sharply, with the result that the seal behaves dynamically. This behavior remains constant from the lowest temperature to more than 3000°C (5430°F). Thus graphite gaskets absorb pressure and temperature load changes, as well as vibrations occurring in the flange.
- The ability of graphite gaskets to conform relatively easily to surface irregularities makes these gaskets particularly suitable for sensitive flanges such as enamel, glass, and graphite flanges.
- Large gaskets and those of complicated shape can be constructed simply from combined segments that overlap. The lapped joints do not constitute weak points.
- Graphite can be used without misgivings in the food industry.

Common gasket constructions include:

- Plain graphite gaskets
- Graphite gaskets with steel-sheet inserts
- Graphite gaskets with steel-sheet inserts and inner or inner and outer edge cladding
- Grooved metal gaskets with graphite facings
- Spiral-wound gaskets

Because of the graphite structure, plain graphite gaskets are sensitive to breakage and surface damage. For this reason, graphite gaskets with steel inserts and spiral-wound gaskets are commonly preferred. There are, however, applications where the unrestrained flexibility of the plain graphite gasket facilitates sealing.

Spiral-Wound Gaskets

Spiral-wound gaskets consist of a V-shaped metal strip which is spirally wound on edge, and a soft filler inlay between the laminations. Several turns of the metal strip at start and finish are spot welded to prevent the gasket from unwinding. The metal strip provides a degree of resiliency to the gasket, which compensates for minor flange movements, whereas the filler material is the sealing medium which flows into the imperfections of the flange face.

Manufacturers specify the amount of compression for the installed gasket to ensure that the gasket is correctly stressed and exhibits the desired resiliency. The resultant gasket operating thickness must be controlled by controlled bolt loading, or the depth of a recess for the gasket in the flange, or by inner and/or outer compression rings. The inner compression ring has the additional duty to protect the gasket from erosion by the fluid, while the outer compression ring locates the gasket within the bolt diameter.

The load-carrying capacity of the gasket at the operating thickness is controlled by the number of strip windings per unit width, referred to as gasket density. Thus, spiral-wound gaskets are tailor-made for the pressure range for which they are intended.

The diametrical clearance for unconfined spiral-wound gaskets between pipe bore and inner gasket diameter, and between outer gasket diameter and diameter of the raised flange face, should be at least 6 mm (1/4 in) If the gasket is wrongly installed and protrudes into the pipe bore or over the raised flange face, the sealing action of the gasket is severely impaired. The diametrical clearance recommended for confined gaskets is 1.5 mm (1/16 in).

The metal windings are commonly made of stainless steel or nickel-base alloys, which are the inventory materials of most manufacturers. The windings may be made also of special materials such as mild steel, copper, or even gold or platinum. In selecting materials for corrosive fluids or high temperatures, the resistance of the material to stress corrosion or intergranular corrosion must be considered. Manufacturers might be able to advise on the selection of the material. Otherwise, References 23 through 27 may be consulted.

The gasket filler material must be selected for fluid compatibility and temperature resistance. Typical filler materials are asbestos paper or compressed asbestos of various types, PTFE, pure graphite, mica with rubber or graphite binder, and ceramic fiber paper. Manufacturers will advise on the field of application of each filler material.

The filler material also affects the sealability of the gasket. Gaskets with asbestos and ceramic paper filler materials require higher seating stresses than gaskets with softer and more impervious filler materials to achieve comparable fluid tightnesses. They also need more care in the selection of the flange surface finish.

In most practical applications, the user must be content with flange face finishes as commercially available. For otherwise identical geometry of the flange sealing surface, however, the surface roughness may vary widely, typically between 3.2 and 12.5 μm Ra (125 and 500 μin. Ra). Optimum sealing has been achieved with a finish described in ANSI B16.5, with the resultant surface finish limited to the 3.2 to 6.3 μm Ra (125 to 250 μin. Ra) range. Surface roughness higher than 6.3 μm Ra (250 μin. Ra) may require unusually high seating stresses to produce the desired flange seal. On the other hand, surface finishes significantly smoother than 3.2 μm Ra (125 μin. Ra) may result in poor sealing performance, probably because of insufficient friction between gasket and flange faces to prevent lateral displacement of the gasket.

A manufacturer's publication dealing with design criteria of spiral-wound gaskets may be found in Reference 27.

Gasket Blowout

Flanged joints with unconfined gaskets should be designed so that gasket blowout is preceded by a leakage warning. The design of the joint must therefore be inspected for conditions of likely gasket blowout and incipient leakage flow.

The condition of safety against blowout is satisfied if the friction force at the gasket faces exceeds the fluid force acting on the gasket in the radial direction, as expressed by the equation:

$$2\mu F \geqslant Pt\pi\, d_m \quad \text{or} \quad F \geqslant \frac{Pt\pi\, d_m}{2\mu} \tag{2-3}$$

where
  μ = friction factor
  F = gasket working load
  P = gauge fluid pressure
  t = gasket thickness
  d_m = mean gasket diameter

The joint begins to leak if:

$$F \geqslant wPm\, \pi\, d_m \tag{2-4}$$

in which
  m = gasket factor
  w = gasket width

The gasket factor is a measure of the sealing ability of the gasket, and it defines the ratio of residual gasket stress to the fluid pressure at which leakage begins to develop. Its value is found experimentally.

It follows thus from Equations 2-3 and 2-4 that the gasket is safe against blowout without prior leakage warning if:

$$w \geqslant \frac{t}{2\,\mu m} \tag{2-5}$$

Krägeloh[7] regarded a gasket factor of 1.0 and a friction factor of 0.1 safe for most practical applications. Based on these factors, the width of the gasket should be not less than five times its thickness to prevent blowout of the gasket without prior leakage warning.

Valve Stem Seals

Compression Packings

Construction. Compression packings are the sealing elements in stuffing boxes (see Figures 3-16 through 3-18). They consist of a soft material which is stuffed into the stuffing box and compressed by a gland to form a seal around the valve stem.

The packings may have to withstand extremes of temperature, be resistant to aggressive media, display a low friction factor and adequate structural strength, and be impervious to the fluid to be sealed. To meet this wide range of requirements, and at the same time offer economy of use, innumerable types of packing constructions have evolved.

Constructions of compression packings for valve stems were, in the past, based largely on asbestos fiber because of its suitability for a wide range of applications. Asbestos is suitable for extremes of temperatures and resistant to a wide range of aggressive media, and does not change its properties over time. On the debit side, asbestos has poor lubricating properties. Therefore a lubricant must be added—one which does not interfere with the properties of asbestos, such as flake graphite or mica. This combination is still permeable to fluids, and a liquid lubricant is added to fill the voids. Again, the lubricant must not interfere with the properties of the construction. This is often very difficult, and in response to this challenge, thousands of variations of packings based on asbestos have been produced.

The types of lubricants used for this purpose are oils and greases when water and aqueous solutions are to be sealed, and soaps and insoluble substances when fluids like oil or gasoline are to be sealed. Unfortunately, liquid lubricants tend to migrate under pressure, particularly at higher temperatures, causing the packing to shrink and harden. Such packings must, therefore, be retightened from time to time to make up for loss of packing volume. To keep this loss to a minimum, the liquid content of valve stem packings is normally held to 10% of the weight of the packing.

With the advent of PTFE, a solid lubricant became available that can be used in fibrous packings without the addition of a liquid lubricant.

Asbestos is now avoided in packings where possible, replaced by polymer filament yarns such as PTFE and aramid, and by pure graphite fiber or foil. Other packing materials include vegetable fibers such as cotton, flax, and ramie (frequently lubricated with PTFE); and twisted and folded metal ribbons.

The types of fibrous packing constructions in order of mechanical strength are loose fill, twisted yarn, braid over twisted core, square-plait braid, and interbraid constructions. The covers of the latter three types of packing constructions often contain metal wire within the strands to increase the mechanical strength of the packing for high fluid pressure and high temperature applications.

Reference 67 offers excellent advice on selection and application of compression packings. Standards on packings may be found in Appendix C.

Sealing action. The sealing action of compression packings is due to their ability to expand laterally against the stem and stuffing box walls when stressed by tightening of the gland.

The stress exerted on the lateral faces of a confined elastic solid by an applied axial stress depends on Poisson's ratio for the material, as expressed by:

$$\sigma_1 = \sigma_a \left(\frac{1 - \mu}{\mu} \right) \qquad (2\text{-}6)$$

where
  σ_1 = lateral stress
  σ_a = axial stress
  μ = Poisson's ratio
    = ratio of lateral expansion to axial compression of an elastic solid compressed between two faces

Thus, the lateral stress equals the axial stress only if $\mu = 0.5$, in which case the material is incompressible in bulk.

A material with a Poisson's ratio nearly equal to 0.5 is soft rubber, and it is known that soft rubber transmits pressure in much the same way as a liquid.[8] Solid PTFE has a Poisson's ratio of 0.46 at 23°C (73°F) and 0.36 at 100°C (212°F).[9] A solid PTFE packing is capable of transmitting 85% and 56% of the axial stress to the lateral faces at the respective temperatures. Other packing materials, however, are much more compressible in bulk, so Poisson's ratio, if it can be defined for these materials, is considerably less than 0.5.

When such packing is compressed in the stuffing box, shrinkage of the packing causes friction between itself and the side walls, which prevents the transmission of the full gland force to the bottom of the packing. This fall in axial packing pressure is quite rapid, and its theoretical value can be calculated.[10,11]

The theoretical pressure distribution, however, applies to static conditions only. When the stem is being moved, a pressure distribution takes place so that an analysis of the actual pressure distribution becomes increasingly difficult.

The pressure distribution is also influenced by the mode of packing installation. If the packing consists of a square cord, bending of the packing around the stem causes the packing to initially assume the shape of a trapezoid. When compressing the packing, the pressure on the inner periphery will be higher than on the outer periphery. Preformed packing rings overcome this effect on the pressure distribution.

When the fluid pressure applied to the bottom of the packing begins to exceed the lateral packing pressure, a gap develops between the packing and the lateral faces, allowing the fluid to enter this space. In the case of low pressure applications, the gland may finally have to be retightened to maintain a fluid seal.

When the fluid pressure is high enough, the sealing action takes place just below the gland, where the fluid pressure attempts to extrude the packing through the gland clearances. At this stage, the sealing action has become automatic.

Readings of the fluid pressure gradient of leakage flow along the stuffing box of rotating shafts, shown in Figure 2-3, confirm this function of the stuffing box seal.[10,11] The pressure gradient at low fluid pressures is more or less uniform, which indicates little influence by the fluid pressure on the sealing action. On the other hand, the readings at high fluid pressures show that 90% of the pressure drop occurs across the packing ring, just below the gland. This indicates a dominant influence of the fluid pressure on the sealing action.

In the case of high fluid pressures, therefore, the packing ring just below the gland is the most important one, and must be selected for resistance to extrusion and wear and be carefully installed. Also, extra long stuffing boxes for high pressure applications do not serve the intended purpose.

If the packing is incompressible in bulk, as in the case of soft rubber, the axial packing pressure introduced by tightening of the gland will produce a uniform lateral packing pressure over the entire length of the packing. Fluid pressure applied to the bottom of the packing increases the lateral packing pressure by the amount of fluid pressure, so the sealing action is automatic once interference between packing and the lateral restraining faces has been established.

Unfortunately, rubber tends to grip the stem and impede its operation unless the inner face of the rubber packing is provided with a slippery surface. For this reason, rubber packings are normally used in the form of O-rings, which because of their size offer only a narrow contact face to the stem.

Corrosion of stainless steel valve stems by packings. Stainless steel valve stems — in particular those made of AISI type 410 (13Cr) steel — corrode frequently where the face contacts the packing. The corrosion

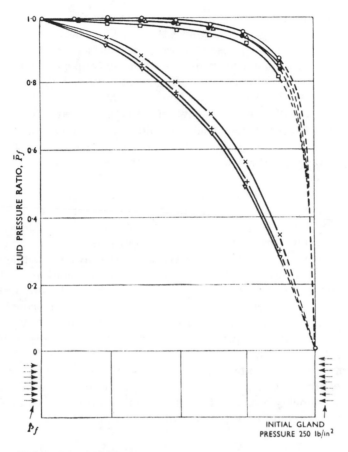

Figure 2-3. Distribution of fluid pressure for four rings of PTFE-impregnated plaited cotton packing.

$\hat{P}_f$ = applied fluid pressure

$\bar{P}_f$ = ratio fo fluid pressure along shaft to applied fluid pressure

Each set of measurements taken 6 hours after change of pressure. Shaft speed: 850 rev/min. Applied gland pressure: 250 lb/in^2. Water pressure, lb/in^2: ○ 1000, △ 700, ● 400, □ 250, × 75, + 26,   2. (Reprinted from *Proceedings of the Institution of Mechanical Engineers,* London, 174 No. 6, 1960, p. 278, by D.F. Denny and D.E. Turnbull.)

occurs usually during storage preceding service, when the packing is saturated with water from the hydrostatic test. If the valve is placed into service immediately after the hydrostatic test, no corrosion occurs.[18] H. J. Reynolds, Jr. has published the results of his investigations into this corrosion phenomenon; the following is an abstract.[12] Corrosion of stainless steel valve stems underlying wet packing is theorized to be the result of the deaerated environment imposed on the steel surface by the restricting packing — an environment which influences the active-passive nature of the metal. Numerous small anodes are created at oxygen-deficient sensitive points of the protective oxide surface film on the stainless steel. These, along with large masses of retained passive metal acting as cathodes, result in galvanic cell action within the metal. Graphite often contained in the packing acts as a cathodic material to the active anodic sites on the steel, and appreciably aggravates the attack at the initial corrosion sites through increased galvanic current density.

Because of the corrosion mechanism involved, it is impractical to make an effective noncorrosive packing using so-called noncorrosive ingredients. Incorporating a corrosion inhibitor into the packing is thus required, which will influence the anodic or cathodic reactions to produce a minimum corrosion rate. Of the anodic inhibitors evaluated, only those containing an oxidizing anion, such as sodium nitrite, are efficient. Cathodic protection by sacrificial metals such as zinc, contained in the packing, also provides good corrosion control. Better protection with a minimum effect on compression and serviceability characteristics of the packing is provided by homogeneously dispersed sodium nitrite and a zinc-dust interlayer incorporated into the material.

High chromium-content stainless steels — especially those containing nickel — exhibit a marked increase in resistance to corrosion by inhibited packing, presumably because of the more rapidly protective oxide surface film and better retention of the passivating film.

Lip-Type Packings

Lip-type packings expand laterally because of the flexibility of their lips, which are forced against the restraining side walls by the fluid pressure. This mode of expansion of the packing permits using relatively rigid construction materials which would not perform as well in compression packings. On the debit side, the sealing action of lip-type packings is in one direction only.

Most lip-type packings for valve stems are made of virgin or filled PTFE. However, fabric-reinforced rubber and leather are also used, mainly for hydraulic applications. Most lip-type packings for valve stems are V-shaped, because they accommodate themselves conveniently in narrow packing spaces.

The rings of V-packings made of PTFE and reinforced rubber are designed to touch each other on small areas near the tips of their lips, and large areas are separated by a gap which permits the fluid pressure to act freely on the lips. Leather V-packing rings lack the rigidity of those made of PTFE and reinforced rubber, and are therefore designed to fully support each other.

V-packings made of PTFE and reinforced rubber are commonly provided with flared lips which automatically preload the restraining lateral faces. In this case, only slight initial tightening of the packing is necessary to achieve a fluid seal. V-packing rings made of leather have straight walls and require a slightly higher axial preload. If a low packing friction is important, as in automatic control valves, the packing is frequently loaded from the bottom by a spring of predetermined strength to prevent manual overloading of the packing.

Squeeze-Type Packings

The name squeeze-type packing applies to O-ring packings and the like. Such packings are installed with lateral squeeze, and rely on the elastic strain of the packing material for the maintenance of the lateral preload. When the fluid pressure enters the packing housing from the bottom, the packing moves towards the gap between the valve stem and the back-up support and thereby plugs the leakage path. When the packing housing is depressurized again, the packing regains its original configuration. Because elastomers display the high yield strain necessary for this mode of action, most squeeze packings are made of these materials.

Extrusion of the packing is controlled by the width of the clearance gap between the stem and the packing back-up support, and by the rigidity of the elastomer as expressed by the modulus of elasticity. Manufacturers express the rigidity of elastomers conventionally in terms of Durometer hardness, although Durometer hardness may express different moduli of elasticity for different classes of compounds. Very small clearance gaps are controlled by leather or plastic back-up rings which fit tightly around the valve stem. Manufacturers of O-ring packings supply tables which relate the Durometer hardness and the clearance gap around the stem to the fluid pressure at which the packing is safe against extrusion.

Thrust Packings

Thrust packings consist of a packing ring or washer mounted between shoulders provided on bonnet and valve stem, whereby the valve stem is free to move in an axial direction against the packing ring. The initial stem seal may be provided either by a supplementary radial packing such as a compression packing, or by a spring which forces the shoulder of the stem against the thrust packing. The fluid pressure then forces the shoulder of the stem into more intimate contact with the packing.

Thrust packings are found frequently in ball valves such as those shown in Figures 3-55 through 3-60.

Diaphragm Valve-Stem Seals

Diaphragm valve-stem seals represent flexible pressure-containing valve covers which link the valve stem with the closure member. Such seals prevent any leakage past the stem to the atmosphere, except in the case of a fracture of the diaphragm. The shape of the diaphragm may represent a dome, as in the valve shown in Figure 3-7,[33] or a bellows, as in the valves shown in Figures 3-6 and 3-38. Depending on the application of the valve, the construction material of the diaphragm may be stainless steel, a plastic, or an elastomer.

Dome-shaped diaphragms offer a large uncompensated area to the fluid pressure, so the valve stem has to overcome a correspondingly high fluid load. This restricts the use of dome-shaped diaphragms to smaller valves, depending on the fluid pressure. Also, because the possible deflection of dome-shaped diaphragms is limited, such diaphragms are suitable only for short-lift valves.

Bellows-shaped diaphragms, on the other hand, offer only a small uncompensated area to the fluid pressure, and therefore transmit a correspondingly lower fluid load to the valve stem. This permits bellows-shaped diaphragms to be used in larger valves. In addition, bellows-shaped diaphragms may be adapted to any valve lift.

To prevent any gross leakage to the atmosphere from a fracture of the diaphragm, valves with diaphragm valve-stem seals are frequently provided with a secondary valve-stem seal such as a compression packing.

Flow Through Valves

Valves may be regarded as analogous to control orifices in which the area of opening is readily adjustable. As such, the friction loss across the valve varies with flow, as expressed by the general relationship:

$$v \propto (\Delta h)^{1/2}$$

$$v \propto (\Delta p)^{1/2}$$

where
   v = flow velocity
   Δh = headloss
   Δp = pressure loss

For any valve position, numerous relationships between flow and flow resistance have been established, using experimentally determined resistance or flow parameters. Common parameters so determined are the resistance coefficient ζ and, dependent on the system of units, the flow coefficients C_v, K_v, and A_v. It is standard practice, to base these parameters on the nominal valve size.[30,65]

Resistance Coefficient ζ

The resistance coefficient ζ defines the friction loss attributable to a valve in a pipeline in terms of velocity head or velocity pressure, as expressed by the equations

$$\Delta h = \zeta \frac{v^2}{2g} \text{ (coherent SI or imperial units)} \tag{2-7}$$

$$\text{and: } \Delta p = \zeta \frac{v^2 \rho}{2} \text{ (coherent SI units)} \tag{2-8}$$

$$\text{or: } \Delta p = \zeta \frac{v^2 \rho}{2g} \text{ (coherent imperial units)} \tag{2-9}$$

where
  ρ = density of fluid
  g = local acceleration due to gravity

The equations are valid for single-phase flow of Newtonian liquids and for both turbulent and laminar flow conditions. They may also be used for flow of gas at low Mach numbers. As the Mach number at the valve inlet approaches 0.2, the effects of compressibility become noticeable but are unlikely to be significant even for Mach numbers up to 0.5.[30]

Valves of the same type but of different manufacture, and also of the same line but different size, are not normally geometrically similar. For this reason, the resistance coefficient of a particular size and type of valve can differ considerably between makes. Table 2-3 can therefore provide only typical resistance-coefficient values. The values apply to fully open valves only and for $Re \geqslant 10^4$. Correction factor K_1 for partial valve opening may be obtained from Figures 2-4 through 2-7.

The Engineering Sciences Data Unit, London,[30] deals more comprehensively with pressure losses in valves. Their publication also covers correction factors for $Re < 10^4$ and shows the influence of valve size on the ζ-value and scatter of data, as obtained from both published and unpublished reports and from results obtained from various manufacturers.

In the case of partially open valves and valves with reduced seat area, as in valves with a converging/diverging flow passage, the energy of the flow stream at the vena contracta converts partially back into static energy.

Table 2-3
Approximate Resistance Coefficients of
Fully Open Valves Under Conditions of
Fully Turbulent Flow.

Globe valve, standard pattern:
- Full bore seat, cast. $\zeta = 4.0$–10.0
- Full bore seat, forged (small sizes only). $\zeta = 5.0$–13.0

Globe valve, 45° oblique pattern:
- Full bore seat, cast. $\zeta = 1.0$–3.0

Globe valve, angle pattern:
- Full bore seat, cast. $\zeta = 2.0$–5.0
- Full bore seat, forged (small sizes only). $\zeta = 1.5$–3.0

Gate valve, full bore: $\zeta = 0.1$–0.3

Ball valve, full bore: $\zeta = 0.1$

Plug valve, rectangular port:
- Full flow area. $\zeta = 0.3$–0.5
- 80% flow area. $\zeta = 0.7$–1.2
- 60% flow area. $\zeta = 0.7$–2.0

Plug valve, circular port, full bore: $\zeta = 0.2$–0.3

Butterfly valve, dependent on blade thickness: $\zeta = 0.2$–1.5

Diaphragm valve:
- Weir type. $\zeta = 2.0$–3.5
- Straight through type. $\zeta = 0.6$–0.9

Lift check valve (as globe valve):

Swing check valve: $\zeta = 1.0$

Tilting-disc check valve: $\zeta = 1.0$

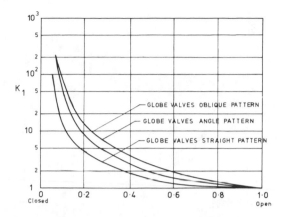

Figure 2-4. Approximate effect of partial opening on resistance coefficient of globe valves. (Courtesy of Engineering Sciences Data Unit. Reproduced from Item No. 69022, Figure 13.)

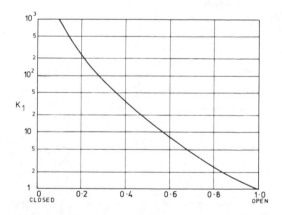

Figure 2-5. Approximate effect of partial opening on resistance coefficient of gate valves. (Courtesy of Engineering Sciences Data Unit. Reproduced from Item No. 69022, Figure 14.)

Figure 2-8 shows the influence of the pressure recovery on the resistance coefficient of fully open venturi-type gate valves in which the gap between the seats is bridged by an eye piece.

The amount of static energy recovered depends on the ratio of the diameters of the flow passage (d^2/D^2), the taper angle ($\alpha/2$) of the diffuser, and the length (L) of straight pipe after the valve throat in terms of pipe diameter (d = valve throat diameter, and D = pipe diameter).

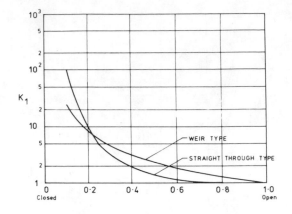

Figure 2-6. Approximate effect of partial opening on resistance coefficient of diaphragm valves. (Courtesy of Engineering Sciences Data Unit. Reproduced from Item No. 69022, Figure 12.)

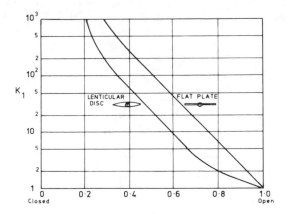

Figure 2-7. Approximate effect of partial opening of resistance coefficient of butterfly valves. (Courtesy of Engineering Sciences Data Unit. Reproduced from Item No. 69022, Figure 15.)

If the straight length of pipe after the valve throat $\geq$ 12 D, the pressure loss cannot exceed the Borda-Carnot loss, which is:

$$\Delta h = \frac{(v_d - v_D)^2}{2g}$$

(2-10)

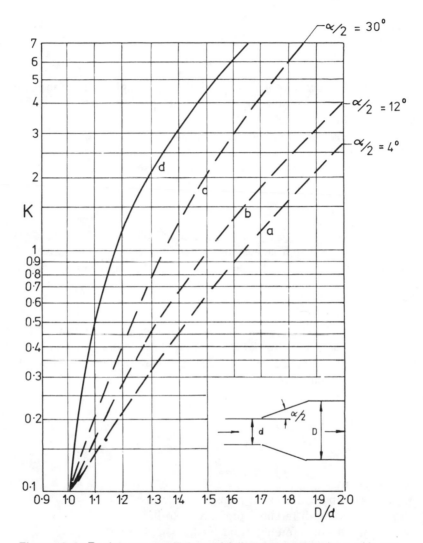

Figure 2-8. Resistance coefficient of fully open gate valves with converging-diverging flow passage and eye piece between the seats.

Curves a to c : L ⩾ 12D.
Curve d   : L = zero.

L = straight length of pipe downstream of venturi throat. (Courtesy of VDI Verlag GmbH, Düsseldorf. Reproduced from BWK Arbeitsblatt 42, Dec. 1953, by H. Haferkamp and A. Kreuz.)

where
v_d = flow velocity in valve throat
v_D = flow velocity in pipeline

This maximum pressure loss occurs if the taper angle ($\alpha/2$) of the diffuser > 30°. With a decreasing taper angle, the pressure loss decreases and reaches its lowest value at a taper angle of 4°.

If the straight pipe after the valve throat < 12 D, the pressure loss increases and reaches its maximum value when the static energy converted into kinetic energy gets completely lost, in which case:

$$\Delta h = \frac{(v_d{}^2 - v_D{}^2)}{2g} \tag{2-11}$$

The friction loss approaches this maximum value if the valve with a converging/diverging flow passage is mounted directly against a header.

Flow Coefficient C_v

The flow coefficient C_v states the flow capacity of a valve in gal (U.S.)/min of water at a temperature of 60°F that will flow through the valve with a pressure loss of one pound per square inch at a specific opening position, as defined by the equation:

$$C_v = Q \left(\frac{\Delta p_0}{\Delta p} \times \frac{\rho}{\rho_0} \right)^{1/2} \tag{2-12}$$

where
Q = gal/min
Δp_0 = reference differential pressure = 1 lb/in.^{2}
Δp = operating differential pressure in lb/in.^{2}
ρ_0 = density of reference fluid, water = 62.4 lb/ft^3
ρ = density of operating fluid in lb/ft^3

Because ρ/ρ_0 = specific gravity and the numerical value of Δp_0 is unity, Equation 2-12 is normally presented in the form:

$$C_v = Q \left(\frac{G}{\Delta p} \right)^{1/2} \tag{2-13}$$

where
G = specific gravity

This relationship is incorporated in the International Standards Association (ISA) standards S.39.1 to S.39.4 and Publication 534-1[64] of the International Electrotechnical Commission (IEC), and applies to single phase and fully turbulent flow of Newtonian liquids. For other flow conditions, refer to the above ISA standards, IEC Publication 534-2 Parts 1 and 2, or Reference 63.

Flow Coefficient K_v

The flow coefficient K_v is a version of coefficient C_v in mixed SI units. It states the number of cubic meters per hour of water at a temperature between 5 and 40°C that will flow through the valve with a pressure loss of one bar at a specific opening position, as defined by the equation:

$$K_v = Q \left(\frac{\Delta p_0}{\Delta p} \times \frac{\rho}{\rho_0} \right)^{1/2} \tag{2-14}$$

where
$\quad Q \quad = m^3/h$
$\quad \Delta p_0 = $ reference differential pressure $= 1$ bar
$\quad \Delta p \ = $ operating differential pressure, bar
$\quad \rho_0 \ \ = $ density of reference fluid (water $= 1,000$ kg/m^3)
$\quad \rho \quad = $ density of operating fluid, kg/m^3

Because $\rho/\rho_0 = $ specific gravity and the numerical value of Δp_0 is unity, Equation 2-14 is normally presented in the form:

$$K_v = Q \left(\frac{G}{\Delta p} \right)^{1/2} \tag{2-15}$$

where
$\quad G = $ specific gravity

This relationship is contained in IEC Publication 534-1[64] and applies to single phase and fully turbulent flow of Newtonian liquids. For other flow conditions, refer to IEC Publication 534-2 Parts 1 and 2, or to manufacturers' catalogs.

Flow Coefficient A_v

The flow coefficient A_v is a version of the flow coefficient K_v in coherent SI units. A_v states the number of cubic meters per second of water at a temperature between 5 and 40°C that will flow through the valve with a pressure loss of one Pascal at a specific opening position, as defined by the equation:

$$A_v = Q \left(\frac{\rho}{\Delta p}\right)^{1/2} \qquad (2\text{-}16)$$

where
  Q = flow rate, m³/s
  p = operational differential pressure, Pa
  ρ = density of Newtonian liquid, kg/m³

A_v is derived from Equation 2-8, which may be presented in the following forms:

$$Q = A \left(\frac{2}{\zeta}\right)^{1/2} \left(\frac{\Delta p}{\rho}\right)^{1/2} \qquad (2\text{-}17a)$$

$$A \left(\frac{2}{\zeta}\right)^{1/2} = Q \left(\frac{\rho}{\Delta p}\right)^{1/2} \qquad (2\text{-}17b)$$

where
  v = flow velocity of fluid, m/s
  A = cross-sectional area, m²

The expression $A \left(\dfrac{2}{\zeta}\right)^{1/2}$ is replaced in Equation 2-16 by a single expression A_v.

This relationship is contained in IEC Publication 534-1[64] and applies to single phase and fully turbulent flow of Newtonian liquids. For other flow conditions, see IEC Publication 534-2 Parts 1 and 2.

Interrelationships Between Resistance and Flow Coefficients

All resistance and flow coefficients are interrelated. If one coefficient is known, all other coefficients can be calculated. These are the interrelationships:

$$\zeta = \frac{889\ d^4}{C_v^2} \qquad C_v = \frac{29.8\ d^2}{\zeta^{1/2}} \qquad (2\text{-}18)$$

$$\zeta = \frac{665\ d^4}{K_v^2} \qquad K_v = \frac{25.8\ d^2}{\zeta^{1/2}} \qquad (2\text{-}19)$$

$$\zeta = \frac{5.13\ d^4}{10^7\ A_v^{\ 2}} \quad A_v = \frac{7.16\ d^2}{10^6\ \zeta^{1/2}} \tag{2-20}$$

$$\frac{K_v}{C_v} = 8.65 \times 10^{-1} \tag{2-21}$$

$$\frac{A_v}{C_v} = 2.40 \times 10^{-5} \tag{2-22}$$

$$\frac{A_v}{K_v} = 2.78 \times 10^{-5} \tag{2-23}$$

Relationship Between Resistance Coefficient and Valve Opening Position

The relationship between the resistance coefficient and valve opening position represents the relationship between flow through the valve and valve opening position only on the basis of constant pressure loss through the valve. The flow characteristic based on this relationship is referred to as the inherent flow characteristic. In the case of automatic control valves, the valve closure member and/or the valve seat orifice are frequently shaped to achieve a particular inherent flow characteristic. Some of the common flow characteristics are shown in Figure 2-9.

In most practical applications, however, the pressure loss through the valve varies with valve opening position. This is illustrated in Figure 2-10 for a flow system incorporating a pump. The upper portion of the figure represents the pump characteristic, displaying flow against pump pressure, and the system characteristic, displaying flow against pipeline

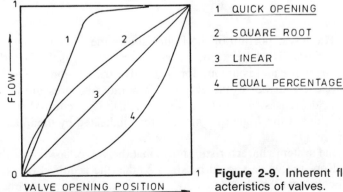

1 QUICK OPENING

2 SQUARE ROOT

3 LINEAR

4 EQUAL PERCENTAGE

Figure 2-9. Inherent flow characteristics of valves.

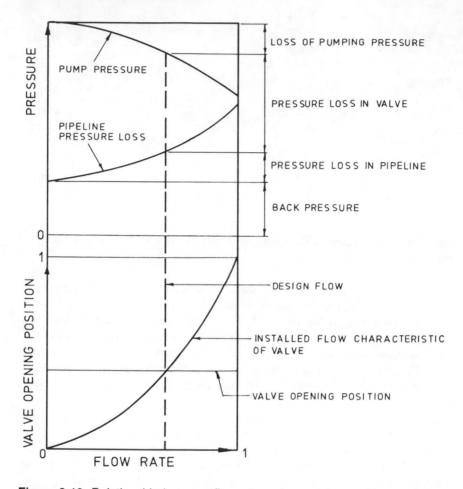

Figure 2-10. Relationship between flow rate, valve opening position, and pressure loss in a pumping system.

pressure loss. The lower portion of the figure shows the flow rate against valve opening position. The latter characteristic is referred to as the *installed valve flow characteristic* and is unique for each valve installation. When the valve is being opened further to increase the flow rate, the pressure at the inlet of the valve decreases, as shown in Figure 2-10. The required rate of valve opening is, therefore, higher in this case than indicated by the inherent flow characteristic.

If the pump and system characteristic shows that the valve has to absorb a high pressure drop, the valve should be sized so that the required pressure drop does not occur near the closed position, since this will promote damage to the seatings from the flowing fluid. This consideration leads frequently to a valve size smaller than the adjoining pipe.

Cavitation in Valves

When a liquid passes through a partially closed valve, the static pressure in the region of increasing velocity and in the wake of the closure member drops and may reach the vapor pressure of the liquid. The liquid in the low-pressure region then begins to vaporize and form vapor-filled cavities which grow around minute gas bubbles and impurities carried by the liquid. When the liquid reaches again a region of high static pressure, the vapor bubbles collapse suddenly or implode. This process is called cavitation.

The impinging of the opposing liquid particles of the collapsing vapor bubble produces locally high but short-lived pressures. If the implosions occur at or near the boundaries of the valve body or the pipe wall, the pressure intensities can match the tensile strength of these parts. The rapid stress reversals on the surface and the pressure shocks in the pores of the boundary surface lead finally to local fatigue failures which cause the boundary surface to roughen until, eventually, quite large cavities form.

The cavitation performance of a valve is typical for a particular valve type, and it is customarily defined by a cavitation index which indicates the degree of cavitation or the tendency of the valve to cavitate. This parameter is presented in the literature in various forms. The following is a convenient index, used by the United States Bureau of Reclamation.[58,59]

$$C = \frac{P_d - P_v}{P_u - P_d} \qquad (2\text{-}15)$$

where
    C = cavitation index
    P_v = vapor pressure relative to atmospheric pressure (negative)
    P_d = pressure in pipe 12 pipe diameters downstream of the valve seat
    P_u = pressure in pipe 3 pipe diameters upstream of the valve seat

Figure 2-11 displays the incipient cavitation characteristics of butterfly, gate, globe, and ball valves, based on water as the flow medium. The characteristics have been compiled by the Sydney Metropolitan Water Sewerage and Drainage Board, and are based on laboratory observations and published data.[13] Because temperature, entrained air, impurities, model tolerances, and the observer's judgment influence the test results, the graphs can serve only as a guide.

The development of cavitation can be minimized by letting the pressure drop occur in stages. The injection of compressed air immediately down-

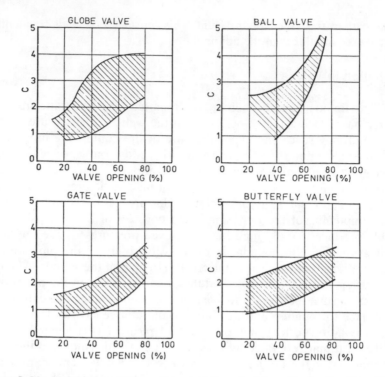

Figure 2-11. Incipient cavitation characteristics of various "in-line" valves. (Courtesy of The Institution of Engineers, Australia.[13])

stream of the valve minimizes the formation of vapor bubbles by raising the ambient pressure. On the debit side, the entrained air will interfere with the reading of any downstream instrumentation.

A sudden enlargement of the flow passage just downstream of the valve seat can protect the boundaries of valve body and pipe from cavitation damage. A chamber with a diameter of 1.5 times the pipe diameter and a length of 8 times the pipe diameter including the exit taper has proved satisfactory for needle valves used in water works.[14]

Waterhammer from Valve Operation

When a valve is being opened or closed to change the flow rate, the change in kinetic energy of the flowing fluid column introduces a transient change in the static pressure in the pipe. In the case of a liquid, this transient change in the static pressure is sometimes accompanied by a shaking of the pipe and a hammering sound — thus the name waterhammer.

The transient pressure change does not occur instantaneously along the entire pipeline but progressively from the point at which the change of flow has been initiated. If, for example, a valve at the end of a pipeline is closed instantaneously, only the liquid elements at the valve feel the valve closure immediately. The kinetic energy stored in the liquid elements then compresses these elements and expands the adjoining pipe walls. The other portion of the liquid column continues to flow at its original velocity until reaching the liquid column which is at rest.

The speed with which the compression zone extends towards the inlet end of the pipeline is uniform and equals the velocity of sound in the liquid within the pipe. When the compression zone has reached the inlet pipe end, all liquid is at rest, but at a pressure above the normal static pressure. The unbalanced pressure now creates a flow in the opposite direction and relieves the rise in the static pressure and the expansion of the pipe wall. When this pressure drop has reached the valve, the whole liquid column is again under normal static pressure, but continues to discharge towards the inlet pipe end so that a wave of subnormal pressure is created, starting at the valve. When this pressure wave has made the round trip, the normal pressure and the original direction of flow are restored. Now the cycle starts again and repeats itself until the kinetic energy of the liquid column is dissipated in friction and other losses.

Joukowsky has shown that instantaneous valve closure raises the static pressure in the pipeline by:

$$\Delta P = av\rho /B \qquad\qquad (2\text{-}16)$$

where

ΔP = rise in pressure above normal

v = velocity of arrested flow

a = velocity of pressure wave propagation

$$= \left[\cfrac{K}{\cfrac{\rho}{B}\left(1 + \cfrac{KDc}{Ee}\right)} \right]^{1/2}$$

ρ = density of the liquid

K = bulk modulus of the liquid

E = Young's modulus of elasticity of the pipe wall material

D = inside diameter of pipe

e = thickness of pipe wall

c = pipe restriction factor ($c = 1.0$ for unrestricted piping)

B (SI units) $= 1.0$

B (imperial units, fps) $= g = 32.174$ ft/s^2

In the case of steel piping with a D/e ratio of 35 and water flow, the pressure wave travels at a velocity of approximately 1200 m/s (about 4000 ft/s), and the static pressure increases by 1.2 MPa for each 1 m/s, or about 50 lb/in^2 for each 1 ft/s instantaneous velocity change.

If the valve does not close instantaneously but within the time of a pressure-wave round trip of 2L/a, where L is the length of the pipeline, the first returning pressure wave cannot cancel the last outgoing pressure wave, and the pressure rise is the same as if the valve were closed instantaneously. This speed of closure is said to be rapid.

If the valve takes longer to close than 2L/a, the returning pressure waves cancel a portion of the outgoing waves so that the maximum pressure rise is reduced. This speed of closure is said to be slow.

If the surge pressure is due to a pump stopping, the calculation of the surge pressure must take into account the pump characteristic and the rate of change of the pump speed after the power supply has been cut off. column. If the valve is still partly open at zero forward velocity, the reversing flow will slam the valve closed. The sudden shut off of the reverse flow then produces a surge pressure, which is in addition to the surge pressure already developing as a result of the forward flow decelerating.

Before the liquid column can begin to retard after the pump has stopped, the pressure wave must travel to the point of reflection. The pressure and elevation at this point then determine the speed at which the flow retards. Thus, if the distance between the check valve and the point of reflection is long, and the elevation and the pressure at this point are low, the system tolerates a slow-closing check valve. On the other hand, if the distance between the check valve and the point of reflection is short, and the pressure at this point is high, the flow reverses almost instantaneously and the check valve must be able to close extremely fast. Such nearly instantaneous reverse flow occurs, for example, in multipump installations in which one pump fails suddenly. Guidelines on the selection of check valves for speed of closure are given in Chapter 4.

Excessively high surge pressures from the operation of stop valves can be avoided by closing the valve more slowly. To achieve a maximum speed of valve closure without producing excessive surge pressure, the valve must be closed so as to produce a uniform rate of change of the flow velocity.

Surge pressures from the closing of check valves in pump installations are not as easily controlled as those from the closing of stop valves. Check valves are operated by the flowing fluid, and their speed of closure is a function of the valve design and the deceleration characteristic of the retarding liquid

Calculation of the fluid pressure and velocity as a function of time, and location along a pipe can be accomplished in several ways. For simple cases, graphical and algebraic methods can be used. However, the ready availability of digital computers has made the use of numerical methods convenient and allows solutions to any desired accuracy to be obtained. See Reference 16 for a description of this calculating method.

In some cases it may be impossible or impractical to reduce the effects of waterhammer by adjusting the valve characteristic. Consideration should then be given to changing the characteristic of the piping system. One of the most common ways of achieving this is to incorporate one or more surge protection devices at strategic locations in the piping system. Such devices may consist of a stand pipe containing gas in direct contact with the liquid or separated from the liquid by a flexible wall, or a pressure relief valve.

The effects of waterhammer may also be altered by deliberately changing the acoustic properties of the fluid. This can be done, for example, by introducing bubbles of a non-dissolvable gas directly into the fluid stream. The effect of this is to reduce the effective density and bulk modulus of the fluid. A similar effect can be achieved if the gas is enclosed in a flexible walled conduit, or hose, which runs the length of the pipe.

If even a small amount of gas is present, the effect of pipe wall elasticity in Equation 2-16 becomes insignificant, and the modified acoustic velocity may be expressed by:

$$a = \sqrt{\frac{BK}{\rho}} \qquad (2\text{-}26)$$

where
  K = modified fluid bulk modulus

$$= \frac{K_1}{1 + \left(\dfrac{V_g}{V_t}\right)\left(\dfrac{K_1}{K_g} - 1\right)}$$

  ρ = modified fluid density

$$= \rho_g \frac{V_g}{V_t} + \rho_1 \frac{V_1}{V_t}$$

  B = 1.0 (coherent SI units)
  B = 32.174 ft/s^2, (imperial units, fps)
  K_g = Bulk modulus of gas
  K_1 = Bulk modulus of liquid
  V_g = Volume of gas
  V_1 = Volume of liquid
  V_t = total volume
  ρ_g = density of gas
  ρ_1 = density of liquid

Figure 2-12 shows the effect of air content on the wave propagation velocity in water. Note that a small amount of air content produces a wave speed

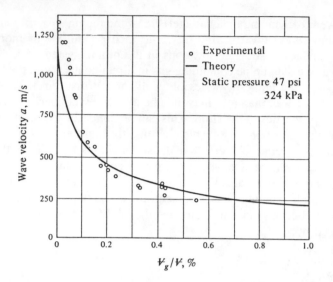

Figure 2-12. Propagation velocity α of a pressure wave in pipeline for varying air content (theoretical and experimental results).[68] Reprint from "Fluid Transients" by E. B. Wylie and V. L. Streeter, by courtesy of the authors.

less than the speed of sound in air, which, for dry air at 20°C (68°F) and atmospheric pressure, is 318 m/s (1042 ft/s).

Attenuation of Valve Noise[38]

The letting down of gas by valves from a high to a low pressure can produce a troublesome and, in extreme cases, unbearable noise. A major portion of the noise arises from the turbulence generated by the high-velocity jet shearing the relatively still medium downstream of the valve. A silencer found successful in combating this noise is the perforated diffuser, in which the gas is made to flow through numerous small orifices. The diffuser may consist of a perforated flat plate, cone, or bucket.

The diffuser attenuates the low and mid frequencies of the valve noise, but also regenerates a high frequency noise in the perforations, which, however, is more readily attenuated by the passage through the pipe and the air than the lower frequencies. A second beneficial effect of the diffuser is to distribute the flow more evenly over the cross section of the pipe.

Ingard has shown that the normalized acoustic resistance of a perforated flat plate mounted across the pipe is directly proportional to both the Mach number of the flow through the perforations and the factor

$$\left(\frac{1 - \sigma}{\sigma} \right)^2$$

where σ = open area ratio of the perforated plate.[25] Although this cannot be directly related to noise attenuation, it would appear that the Mach number should be as large as possible, and σ as small as possible. For practical purposes, a maximum Mach number of 0.9 is suggested. If the available pressure drop across the diffuser is limited, a Mach number with a lower value may have to be chosen. Practical values for the open area ratio may be taken as between 0.1 and 0.3. Values lower than 0.1 may result in an excessively large diffuser, while values higher than 0.3 may result in too low an attenuation.

The peak frequency of the jet noise is also inversely proportional to the diameter of the jet. Therefore, from the point of noise attenuation, the diameter of the perforations should be as small as possible. To avoid the nozzles from becoming blocked, nozzles with a minimum diameter of 5 mm are frequently used.

If the flow velocity in the pipe downstream of the silencer is high, the boundary layer turbulence along the pipe may generate a noise comparable with the attenuated valve noise. Experience suggests that this will not be a problem if the Mach number of the flow in the pipe is kept below about 0.3.

Predicting valve noise and silencer performance is a complex matter. Discussions on these subjects, including the design of silencers, may be found in the References 45 through 49. Further discussions on the generation and radiation of piping noise may be found in References 51 and 52.

Equations covering the sizing of the diffuser and the exhaust stack, and sample calculations, may be found in Chapter 5, pages 216 and 222. The derivation of these equations is given in Appendix A.

3

Manual Valves

Functions of Manual Valves

Manual valves serve three major functions in fluid-handling systems: stopping and starting flow, controlling flow rate, and diverting flow. Valves for stopping and starting flow are frequently employed also for controlling flow rate, and vice versa, while valves for diverting flow are designed for that single purpose.

Grouping of Valves by Method of Flow Regulation

Manual valves may be grouped according to the way the closure member moves onto the seat. Four groups of valves are thereby distinguishable:

1. Closing-down valves. A stopper-like closure member is moved to and from the seat in the direction of the seat axis.
2. Slide valves. A gate-like closure member is moved across the flow passage.
3. Rotary valves. A plug or disc-like closure member is rotated within the flow passage, around an axis normal to the flow stream.
4. Flex-body valves. The closure member flexes the valve body.

Each valve group represents a number of distinct types of valves which use the same method of flow regulation, but differ in the shape of the closure member. For example, plug valves and butterfly valves are both rotary valves, but of a different type. In addition, each valve is made in numerous variations to satisfy particular service needs. Figure 3-1 illustrates the principal methods of flow regulation and names the types of valves which belong to a particular valve group.

Groups of Valves By Method of Flow Regulation	Valve Types	Refer to Page
Closing Down 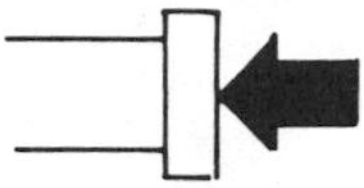	Globe Valve	48
	Piston Valve	63
Sliding 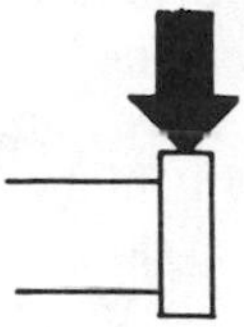	Parallel Gate Valve	67
	Wedge Gate Valve	76
Rotating 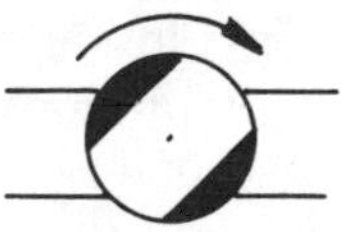	Plug Valve	89
	Ball Valve	98
	Butterfly Valve	109
Flexing of Valve Body 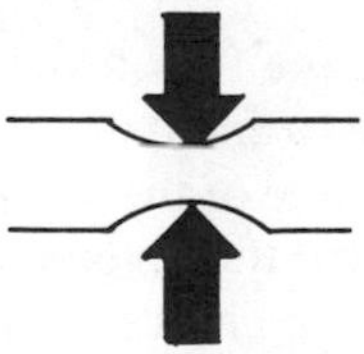	Pinch Valve	126
	Diaphragm Valve	131

Figure 3-1. Principal types of valves grouped according to method of flow regulation.

Selection of Valves

The method by which the closure member regulates the flow and the configuration of the flow path through the valve impart a certain flow characteristic to the valve, which is taken into account when selecting a valve for a given flow-regulating duty.

Valves for Stopping and Starting Flow

These valves are normally selected for low flow resistance, as provided by valves with a straight-through flow passage. Such valves are slide valves, rotary valves, and flex-body valves. Closing-down valves offer by their tortuous flow passage a higher flow resistance than other valves and are therefore less frequently used for this purpose. However, if the higher flow resistance can be accepted — as is frequently the case — closing-down valves may likewise be used for this duty.

Valves for Control of Flow Rate

These are selected for easy adjustment of the flow rate. Closing-down valves lend themselves for this duty because of the directly proportional relationship between the size of the seat opening and the travel of the closure member. Rotary valves and flex body valves also offer good throttling control, but normally only over a restricted valve-opening range. Gate valves, in which a circular disc travels across a circular seat opening, achieve good flow control only near the closed valve position and are therefore not normally used for this duty.

Valves for Diverting Flow

These valves have three or more ports, depending on the flow diversion duty. Valves which adapt readily to this duty are plug valves and ball valves. For this reason, most valves for the diversion of flow are one of these types. However, other types of valves have also been adapted for the diversion of flow, in some cases by combining two or more valves which are suitably interlinked.

Valves for Fluids with Solids in Suspension

If the fluid carries solids in suspension, the valves best suited for this duty have a closure member which slides across the seat with a wiping motion. Valves in which the closure member moves squarely on and off the seat may trap solids and are therefore suitable only for essentially clean fluids unless the seating material can embed trapped solids.

Valve End Connections

Valves may be provided with any type of end connection used to connect piping. The most important of these for valves are threaded, flanged, and welding end connections.

Threaded end connections. These are made, as a rule, with taper or parallel female threads which screw over tapered male pipe threads. Because a joint made up in this way contains large leakage passages, a sealant or filler is used to close the leakage passages. If the construction material of the valve body is weldable, screwed joints may also be seal welded. If the mating parts of the joint are made of different but weldable materials with widely differing coefficients of expansion, and if the operating temperature cycles within wide limits, seal welding the screwed joint may be necessary.

Valves with threaded ends are primarily used in sizes up to DN 50 (NPS 2). As the size of the valve increases, installing and sealing the joint become rapidly more difficult. Threaded end valves are available, though, in sizes up to DN 150 (NPS 6).

To facilitate the erection and removal of threaded end valves, couplings are used at appropriate points in the piping system. Couplings up to DN 50 (NPS 2) consist of unions in which a parallel thread nut draws two coupling halves together. Larger couplings are flanged.

Codes may restrict the use of threaded end valves, depending on application.

Flanged end connections. These permit valves to be easily installed and removed from the pipeline. However, flanged valves are bulkier than threaded end valves and correspondingly dearer. Because flanged joints are tightened by a number of bolts which individually require less tightening torque than a corresponding screwed joint, they can be adapted for all sizes and pressures. At temperatures above 350°C (660°F), however, creep relaxation of the bolts, gasket, and flanges can, in time, noticeably lower the bolt load. Highly stressed flanged joints can develop leakage problems at these temperatures.

Flange standards may offer a variety of flange face designs and also recommend the appropriate flange face finish. As a rule, a serrated flange face finish gives good results for soft gaskets. Metallic gaskets require a finer flange face finish for best results. Chapter 2, page 12, discusses the design of gaskets.

Welding end connections. These are suitable for all pressures and temperatures, and are considerably more reliable at elevated temperatures and other severe applications than flanged connections. However, removal and re-erection of welding end valves is more difficult. The use of welding end valves is therefore normally restricted to applications in which the valve is expected to operate reliably for long periods, or applications which are critical or which involve high temperatures.

Welding end valves up to DN 50 (NPS 2) are usually provided with welding sockets which receive plain end pipes. Because socket weld joints form a crevice between socket and pipe, there is the possibility of crevice corrosion with some fluids. Also, pipe vibrations can fatigue the joint. Codes restrict, therefore, the use of welding sockets.

Standards Pertaining to Valve Ends

A list of the most important USA and British standards pertaining to valve ends may be found in Appendix C.

Valve Ratings

The rating of valves defines the pressure-temperature relationship within which the valve may be operated.

The responsibility for determining valve ratings has been left over the years largely to the individual manufacturer. The frequent USA practice of stating the pressure rating of general purpose valves in terms of WOG (water, oil, gas) and WSP (wet steam pressure) is a carry-over from the days when water, oil, gas, and wet steam were the substances generally carried in piping systems. The WOG rating refers to the room-temperature rating, while the WSP rating is usually the high-temperature rating. When both a high and a low temperature rating is given, it is generally understood that a straight line pressure-temperature relationship exists between the two points.

Some USA and British standards on flanged valves set ratings which equal the standard flange rating. Both groups of standards specify also the permissible construction material for the pressure-containing valve parts. The rating of welding end valves corresponds frequently to the rating of flanged valves. However, standards may permit welding end valves to be designed to special ratings which meet the actual operating conditions.[44] If the valve contains components made of polymeric materials, the pressure-temperature relationship is limited, as determined by the properties of the polymeric material. Some standards for valves containing such materials — like ball valves — specify a minimum pressure-temperature relationship for the valve. Where such standards do not exist, it is the manufacturer's responsibility to state the pressure and temperature limitations of the valve.

Valve Standards and Standard Organizations[4]

Valve standards are designed to ensure interchangeability and reasonable functioning of the valve. As a rule, valve standards cover face-to-face dimensions, materials of construction, pressure-temperature ratings, design dimensions for some valve components to ensure adequate strength, and procedures for the testing of valves. Detail design is the responsibility of the valve manufacturer.

USA standards exist on industry and national levels. The Manufacturers Standardization Society of the Valve and Fittings Industry (MSS), the American Petroleum Institute (API), and the American Water Works Association (AWWA) are examples of the many industry groups which develop industry standards. When industry standards become stable, they are frequently adopted as national standards under the procedures of organizations such as the American National Standards Institute (ANSI). The MSS withdraws a standard once it is adopted by a recognized standard organization.

British standards are sponsored solely by the British Standards Institution (BSI), an independent body supported by government and industry. User organizations, the valve makers' organization, government departments, and scientific organizations supervise the establishment of valve standards at the invitation of the British Standards Institution. The standards may be designed thereby for use in one particular industry or for general use.

Valve Selection Chart

The valve selection chart shown in Table 3-1 is based on the foregoing guidelines and may be used to select the valve for a given flow-regulating duty. The construction material of the valve is determined on one hand by the operating pressure and temperature in conjunction with the applicable valve standard, and on the other hand by the properties of the fluid, such as its corrosive and erosive properties. The selection of the seating material is discussed in Chapter 2, page 9. The best approach is to consult the valve manufacturer's catalog, which usually states the standard construction material for the valve for given operating conditions. However, it is the purchaser's responsibility to ensure that the construction material of the valve is compatible with the fluid to be handled by the valve.

If the cost of the valve initially selected is too high for the purpose of the valve, the suitability of other types of valves must be investigated. Sometimes a compromise has to be made.

This selection procedure requires knowledge of the available valve types and their variations, and possibly knowledge of the available valve standards. This information is given in the following sections.

Table 3-1
Valve Selection Chart

Valve		Mode of Flow Regulation			Fluid				
Group	Type	On–Off	Throttling	Diverting	Free of solids	Solids in Suspension		Sticky	Sanitary
						non–abrasive	abrasive		
Closing down	Globe:								
	— straight pattern	Yes	Yes		Yes				
	— angle pattern	Yes	Yes		Yes	special	special		
	— oblique pattern	Yes	Yes		Yes	special			
	— multiport pattern			Yes	Yes				
	Piston	Yes	Yes		Yes	Yes	special		
Sliding	Parallel gate:								
	— conventional	Yes			Yes				
	— conduit gate	Yes			Yes	Yes	Yes		
	— knife gate	Yes	special		Yes	Yes	Yes		
	Wedge gate:								
	— with bottom cavity	Yes			Yes				
	— without bottom cavity (rubber seated)	Yes	moderate		Yes	Yes			
Rotating	Plug:								
	— non–lubricated	Yes	moderate	Yes	Yes	Yes			Yes
	— lubricated	Yes		Yes	Yes	Yes	Yes		
	— eccentric plug	Yes	moderate		Yes	Yes		Yes	
	— lift plug	Yes		Yes	Yes	Yes		Yes	
	Ball	Yes	moderate	Yes	Yes	Yes			
	Butterfly	Yes	Yes	special	Yes	Yes			Yes
Flexing	Pinch	Yes	Yes	special	Yes	Yes	Yes	Yes	Yes
	Diaphragm:								
	— weir type	Yes	Yes		Yes	Yes		Yes	Yes
	— straight–through	Yes	moderate		Yes	Yes		Yes	Yes

GLOBE VALVES

Globe valves are closing-down valves in which the closure member is moved squarely on and off the seat. It is customary to refer to the closure member as a disc, irrespective of its shape.

By this mode of disc travel, the seat opening varies in direct proportion to the travel of the disc. This proportional relationship between valve opening and disc travel is ideally suited for duties involving regulation of flow rate. In addition, the seating load of globe valves can be positively controlled by a screwed stem, and the disc moves with little or no friction onto the seat, depending on the design of seat and disc. The sealing capacity of these valves is therefore potentially high. On the debit side, the seatings may trap solids which travel in the flowing fluid.

Globe valves may, of course, be used also for on-off duty, provided the flow resistance from the tortuous flow passage of these valves can be accepted. Some globe valves are also designed for low flow resistance for use in on-off duty. Also, if the valve has to be opened and closed frequently, globe valves are ideally suited because of the short travel of the disc between the open and closed positions, and the inherent robustness of the seatings to the opening and closing movements.

Globe valves may therefore be used for most duties encountered in fluid-handling systems. This wide range of duties has led to the development of numerous variations of globe valves designed to meet a particular duty at the lowest cost. The valves shown in Figures 3-2 through 3-10 are representative of the many variations which are commonly used in pipelines for the control of flow. Those shown in Figures 3-11 through 3-13 are specialty valves which are designed to meet a special duty, as described in the cutlines of these illustrations.

An inspection of these illustrations shows numerous variations in design detail. These are discussed in the following section.

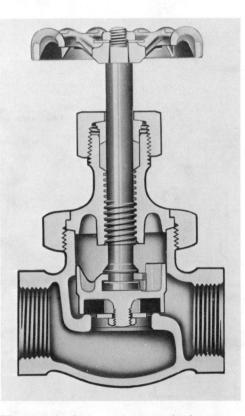

Figure 3-2. Globe valve, standard pattern, union bonnet, internal screw, renewable soft disc. (Courtesy of Crane Co.)

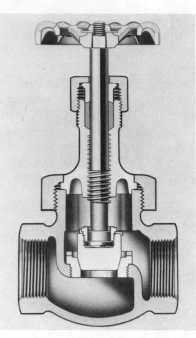

Figure 3-3. Glove valve, standard pattern, union bonnet, internal screw, plug disc. (Courtesy of Crane Co.)

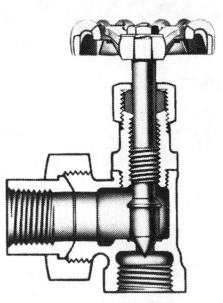

Figure 3-4. Glove valve, angle pattern, screwed-in bonnet, internal screw, needle disc. (Courtesy of Crane Co.)

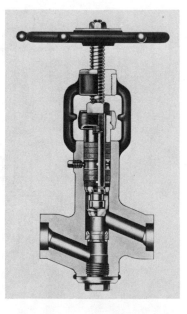

Figure 3-5. Glove valve, standard pattern, integral bonnet, external screw, plug disc. (Courtesy of Velan Engineering Limited.)

Figure 3-6. Globe valve, standard pattern, welded bonnet, external screw, plug disc, bellows stem seal with auxiliary compression packing. (Courtesy of Pegler Hatersley Limited.)

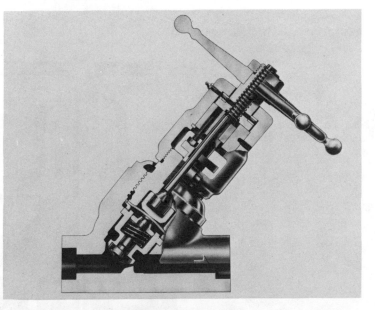

Figure 3-7. Glove valve, oblique pattern, screwed-in and seal-welded bonnet, external screw, plug disc, with domed diaphragm stem seal and auxiliary compression packing for nuclear applications. (Courtesy of Flow Control Division, Rockwell International.)

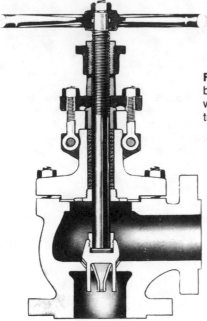

Figure 3-8. Globe valve, angle pattern, bolted bonnet, external screw, plug disc with V-port skirt for sensitive throttling control. (Courtesy of Crane Co.)

Figure 3-9. Globe valve, standard pattern, bolted bonnet, external screw, plug disc. (Courtesy of Crane Co.)

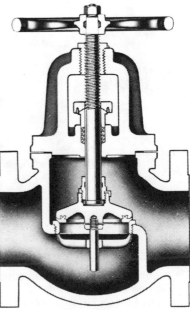

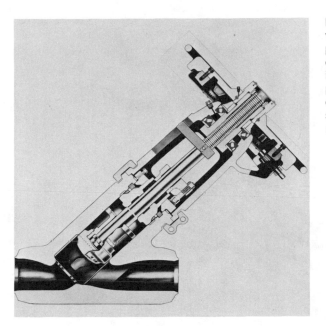

Figure 3-10. Globe valve, oblique pattern, pressure-seal bonnet, external screw, with impact handwheel, plug disc. (Courtesy of Flow Control Division, Rockwell International.)

Figure 3-10a. Globe valve, standard pattern, integral bonnet, plug-type disc integral with non-rotating stem. (Courtesy of Sempell Armaturen.)

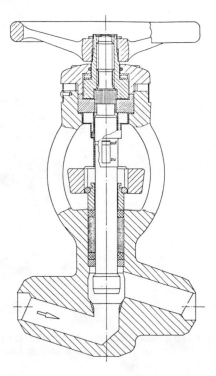

Figure 3-11. Globe valve, oblique pattern, split body, external screw, with seat-wiping mechanism for application in slurry service. (Courtesy of Langley Alloys Limited.)

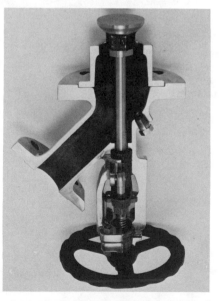

Figure 3-12. Globe valve, adapted for the draining of vessels, seat flush with bottom of vessel. (Courtesy of Langley Alloys Limited.)

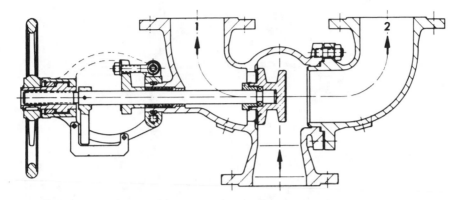

Figure 3-13. Globe valve, three-way, used as change-over valve in pressure relief valve installations: one pressure relief valve is isolated while the second one is in service. (Courtesy of Bopp & Reuther GmbH.)

Valve Body Patterns

The basic patterns of globe-valve bodies are the standard pattern, as in the valves shown in Figures 3-2, 3-3, 3-5, 3-6, and 3-9; the angle pattern, as in the valves shown in Figures 3-4 and 3-8; and the oblique pattern, as in the valves shown in Figure 3-7 and Figures 3-10 through 3-12.

The standard pattern valve body is the most common one, but offers by its tortuous flow passage the highest resistance to flow of the patterns available.

If the valve is to be mounted near a pipe bend, the angle pattern valve body offers two advantages. First, the angle pattern body has a greatly reduced flow resistance compared to the standard pattern body. Second, the angle pattern body reduces the number of pipe joints and saves a pipe elbow.

The oblique pattern globe-valve body is designed to reduce the flow resistance of the valve to a minimum. This is particularly well-achieved in the valve shown in Figure 3-10. This valve combines low flow resistance for on-off duty with the robustness of globe-valve seatings.

Valve Seatings

Globe valves may be provided with either metal seatings or soft seatings.

In the case of metal seatings, the seating stress must not only be high but also circumferentially uniform to achieve the desired degree of fluid tightness. These requirements have led to a number of seating designs; the ones shown in Figure 3-14 are common variations.

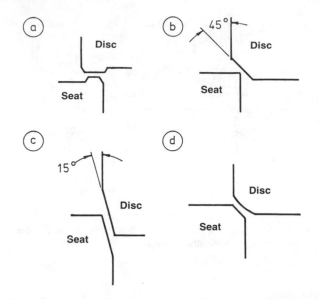

Figure 3-14. Seating configurations frequently employed in globe valves.

Flat seatings (see Figure 3-14a) have the advantage over other types of seatings in that they align readily to each other without having to rely on close guiding of the disc. Also, if the disc is moved onto the seat without being rotated, the seatings mate without friction. The resistance of the seating material to galling is therefore unimportant in this case. Deformation of the roundness of the seat due to pipeline stresses does not interfere with the sealability of the seatings as long as the seat face remains flat. If flow is directed from above the seat, the seating faces are protected from the direct impact of solids or liquid droplets traveling in the fluid.

By tapering the seatings, as shown in Figures 3-14b, c, and d, the seating stress for a given seating load can be greatly increased. However, the seating load can be translated into higher uniform seating stress only if the seatings are perfectly mated; i.e., they must not be mated with the disc in a cocked position. Thus, tapered discs must be properly guided into the seat. Also, the faces of seat and disc must be perfectly round. Such roundness is sometimes difficult to maintain in larger valves where pipeline stresses may be able to distort the seat roundness. Furthermore, as the seatings are tightened, the disc moves further into the seat. Tapered seatings therefore tighten under friction even if the disc is lowered into the seat without being rotated. Thus the construction material for seat and disc must be resistant to galling in this case.

The tapered seatings shown in Figure 3-14b have a narrow contact face, so the seating stress is particularly high for a given seating load. However, the narrow seat face is not capable of guiding the disc squarely into the seat to achieve maximum sealing performance. But if the disc is properly guided, such seatings can achieve an extremely high degree of fluid tightness. On the debit side, narrow-faced seatings are more readily damaged by solids or liquid droplets than wide-faced seatings, so they are used mainly for gases free of solids and liquid droplets.

To improve the robustness of tapered seatings without sacrificing seating stress, the seatings shown in Figure 3-14c are tapered 15° and provided with wide faces which more readily guide the disc into the seat. To achieve a high seating stress, the seat face in initial contact with the disc is relatively narrow, about 3 mm (1/8 in.) wide. The remainder of the seat is tapered slightly steeper. As the seating load increases, the disc slips deeper into the seat, thereby increasing the seating width. Seatings designed in this way are not as readily damaged by erosion as the seatings in Figure 3-14b. In addition, the long taper of the disc improves the throttling characteristic of the valve.

The performance of such seatings may be improved by hollowing out the disc to impart some elasticity to the disc shell, as is done in the valve shown in Figure 3-10a. This elasticity permits the disc to adapt more readily to deviations of the seatings from roundness.

The seatings shown in Figure 3-14d are ball shaped at the disc and tapered at the seat. The disc can therefore roll, to some extent, on the seat until seat and disc are aligned. Because the contact between the seatings approaches that of a line, the seating stress is very high. On the debit side, the line contact is prone to damage from erosion. The ball-shaped seatings are therefore used only for dry gases, which are also free of solids. This construction is used mainly by USA manufacturers.

If the valve is required for fine throttling duty, the disc is frequently provided with a needle-shaped extension, as in the valve shown in Figure 3-4; or with a V-port skirt, as in the valve shown in Figure 3-8; and in the seatings shown in Figure 3-15. In the latter design, the seating faces separate before the V-ports open. The seating faces are, in this way, largely protected against erosion.

An example of soft seating design is the valve shown in Figure 3-2. The soft seating insert is carried in this case by the disc, and may be renewed readily.

Connection of Disc to Stem

The stem of a globe valve may be designed to rotate while raising or lowering the disc, or be prevented from rotating while carrying out this task. These modes of stem operation have a bearing on the design of the disc-to-stem connection.

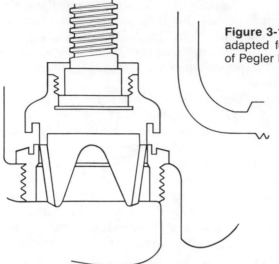

Figure 3-15. Seatings of globe valves adapted for throttling duty. (Courtesy of Pegler Hattersley Limited.)

Most globe valves incorporate a rotating stem because of simplicity of design. If the disc is an integral component of the stem in this case, as it frequently is in small needle valves such as those shown in Figure 3-4, the seatings will mate while the disc rotates, possibly resulting in severe wear of the seatings. Therefore, the main field of application of such valves is for regulating duty with infrequent shut-off duty. For all other duties involving rotating stems, the disc is designed to swivel freely on the stem. However, swivel discs should have minimum free axial play on the stem to prevent the possibility of rapid reciprocating movements of the disc on the stem in the near-closed valve position. Also, if the disc is guided by the stem, there should be little lateral play between stem and disc to prevent the disc from landing on the seat in a cocked position.

In the case of non-rotating stems, as in the valves shown in Figures 3-6, 3-7, 3-10, and 3-10a, the disc may be either an integral part of the stem (see Figure 3-10a) or a separate component from the stem (see Figures 3-6, 3-7, and 3-10). Non-rotating stems are required in valves with diaphragm or bellows valve stem seal, as in Figures 3-6 and 3-7. They are also used in high pressure valves such as those shown in Figures 3-10 and 3-10a to facilitate the erection of power operators.

Inside and Outside Stem Screw

The screw for raising or lowering the stem may be located inside the valve body, as in the valves shown in Figures 3-2 through 3-4, or outside the valve body, as in the valves shown in Figures 3-5 through 3-13.

The inside screw permits an economical bonnet construction, but it has the disadvantage that it cannot be serviced from the outside. This construction is therefore best suited for fluids which have good lubricity. For the majority of minor duties, however, the inside screw gives good service.

The outside screw can be serviced from the outside and is therefore preferred for severe duties.

Bonnet Joints

Bonnets may be joined to the valve body by screwing, flanging, welding, or by means of a pressure-seal mechanism; or the bonnet may be an integral part of the valve body.

The screwed-in bonnet found in the valve shown in Figure 3-4 is one of the simplest and least expensive constructions. However, the bonnet gasket must accommodate itself to rotating faces, and frequent unscrewing of the bonnet may damage the joint faces. Also, the torque required to tighten the bonnet joint becomes very large for the larger valves. For this reason, the use of screwed-in bonnets is normally restricted to valve sizes not greater than ND 80 (NPS 3).

If the bonnet is made of a weldable material, the screwed-in bonnet may be seal welded, as in the valves shown in Figures 3-6 and 3-7, or the bonnet connection may be made entirely by welding. These constructions are not only economical but also most reliable irrespective of size, operating pressure, and temperature. On the debit side, access to the valve internals can be gained only by removing the weld. For this reason, welded bonnets are normally used only where the valve can be expected to be maintenance free for long periods, where the valve is a throw-away valve, or where the sealing reliability of the bonnet joint outweighs the difficulty of gaining access to the valve internals.

The bonnet may also be held to the valve body by a separate screwed union ring, as in the valves shown in Figures 3-2 and 3-3. This construction has the advantage of preventing any motion between the joint faces as the joint is being tightened. Repeatedly unscrewing the bonnet, therefore, cannot readily harm the joint faces. As with the screwed-in bonnet, the use of bonnets with a screwed union ring is restricted to valve sizes normally not greater than DN 80 (NPS 3).

Flanged bonnet joints such as those found in the valves shown in Figures 3-8 and 3-9 have the advantage over screwed joints in that the tightening effort can be spread over a number of bolts. Flanged joints may therefore be designed for any valve size and operating pressure. However, as the valve size and operating pressure increase, the flanged joint becomes increasingly heavy and bulky. Also, at temperatures above 350°C (660°F), creep relaxation can, in time, noticeably lower the bolt load. If the application is critical, the flanged joint may be seal welded.

The pressure-seal bonnet found in the valve shown in Figure 3-10 overcomes this weight disadvantage by letting the fluid pressure tighten the joint. The bonnet seal therefore becomes tighter as the fluid pressure increases. This construction principle is frequently preferred for large valves operating at high pressures and temperatures.

Small globe valves may avoid the bonnet joint altogether, as in the valve shown in Figures 3-5 and 3-10a. Access to the valve internals is through the gland opening, which is large enough to pass the valve components.

Stuffing Boxes and Back Seating

Figures 3-16, 3-17, and 3-18 show three types of stuffing boxes which are typical for valves with a rising stem.

The stuffing box shown in Figure 3-16 is the basic type in which an annular chamber contains the packing between the gland at the top and a shoulder at the bottom. The underside of the stuffing box carries a back seat which, in conjunction with a corresponding seat around the stem, is used to isolate the packing from the system fluid when the valve is fully open.

The stuffing box shown in Figure 3-17 is supplemented with a condensing chamber at the bottom. The condensing chamber originally served as a cooling chamber for condensable gases such as steam. In this particular case, the condensing chamber is provided with a test plug which may be removed to test the back seat for leak tightness.

A third variation of the stuffing box has a lantern ring mounted between two packing sections, as shown in Figure 3-18. The lantern ring is used mainly in conjunction with compression packings and may serve four different purposes:

1. As an injection chamber for a sealant or an extruded or leached-out lubricant.
2. As a pressure chamber in which an external fluid is pressurized to a pressure equal to or slightly higher than the system pressure to prevent any leakage of the system fluid to the outside. The external fluid must thereby be compatible with the system fluid and harmless to the surroundings of the valve.
3. As a sealant chamber in vacuum service into which an external fluid is fed to serve as a sealant.
4. As a leakage collection chamber from which the leakage is piped to a safe location.

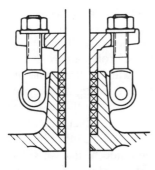

Figure 3-16. Basic stuffing box. (Courtesy of Babcock-Persta Armaturen-Vertriebsgesellschaft mbH.)

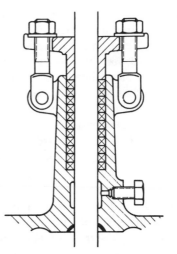

Figure 3-17. Stuffing box with condensing chamber. (Courtesy of Babcock-Persta Armaturen-Vertriebsgesellschaft mbH.)

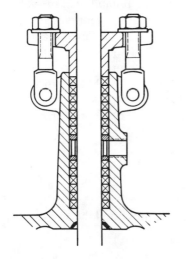

Figure 3-18. Stuffing box with lantern ring. (Courtesy of Babcock-Persta Armaturen-Vertriebsgesellschaft mbH.)

Because the transmission of the gland force to the bottom of the packing ring is poor, the packing stress below the lantern ring is correspondingly lower than directly below the gland. When using the lantern ring for purpose four, the sealing performance of the packing is therefore only limited.[42] In the case of purpose two, the lantern ring has also been substituted by a spring to stress the lower packing section directly, giving excellent results.[57]

Direction of Flow Through Globe Valves

The question of direction of flow through globe valves has two answers.

If the possibility exists that flow from above the disc can remove either the disc from the stem or a component from the disc, flow directed from below the disc is mandatory. In this case, hand-operated globe valves with rotating stem and metal seatings can be closed fluid tight without undue effort, only if the fluid load on the underside of the disc does not exceed about 40-60 kN (9,000–13,000 lb).[60] With a non-rotating stem and roller-bearing supported stem nut, as in the valves shown in Figures 3-10 and 3-10a, hand-operated globe valves with metal seatings may be closed fluid-tight against a fluid load of about 70–100 kN (16,000–22,000 lb), depending on the leakage criterion and the construction of valve.[60] One particular advantage of flow directed from below the disc is that the stuffing box of the closed valve is relieved from the upstream pressure. On the debit side, if the valve has been closed against a hot fluid such as steam, thermal contraction of the stem after the valve has been closed can be just enough to induce seat leakage.

If flow is directed from above the disc, the closing force from the fluid acting on top of the disc supplements the closing force from the stem. Thus, this direction of flow increases greatly the sealing reliability of the valve. In this case, hand-operated globe valves with a rotating stem may be opened without excessive effort, only if the fluid load acting on top of the disc does not exceed about 40-60 kN (9000-13000 lb).[60] If the stem is of the non-rotating type with a roller-bearing supported stem nut, the globe valve may be opened by hand against a fluid load of about 70-100 kN (16000-22000 lb).[60] If the fluid load on top of the disc is higher, a bypass valve may have to be provided which permits the downstream system to be pressurized before the globe valve is opened.

Standards Pertaining to Globe Valves

A list of USA and British standards pertaining to globe valves may be found in Appendix C.

Applications

Duty:
   Controlling flow.
   Stopping and starting flow.
   Frequent valve operation.
Service:
   Gases essentially free of solids.
   Liquids essentially free of solids.
   Vacuum.
   Cryogenic.

PISTON VALVES

Piston valves are closing-down valves in which a piston-shaped closure member intrudes into or withdraws from the seat bore, as in the valves shown in Figures 3-19 through 3-23.

In these valves, the seat seal is achieved between the lateral faces of the piston and the seat bore. When the valve is being opened, flow cannot start until the piston has been completely withdrawn from the seat bore. Any erosive damage occurs, therefore, away from the seating surfaces. When the valve is being closed, the piston tends to wipe away any solids which might have deposited themselves on the seat. Piston valves may thus handle fluids which carry solids in suspension. When some damage occurs to the seatings, the piston and the seat can be replaced *in situ*, and the valve is like new without any machining.

Like globe valves, piston valves permit good flow control. If sensitive flow adjustment is required, the piston may be fitted with a needle-shaped extension. Piston valves are also used for stopping and starting flow when flow resistance due to the tortuous flow passage is accepted.

Construction

The seatings of piston valves are formed by the lateral faces of the valve bore and the piston. A fluid-tight contact between these faces is achieved by a packing which either forms part of the valve bore, as in the valves shown in Figures 3-19 through 3-21, or part of the piston, as in the valves shown in Figures 3-22 and 3-23. Packings commonly used for this purpose are compression packings based on compressed asbestos or PTFE, and O-ring packings.

In the case of the piston valve shown in Figure 3-19, the piston moves in two packings which are separated by a lantern ring. The lower packing represents the seat packing, while the upper packing seals the piston to the atmosphere. The bonnet serves thereby as the gland which permits both packings to be tightened through tightening of the cover bolts. Disc springs under the nuts of the cover bolts minimize variations in packing stress due to thermal contraction and expansion of the valve parts. When one of the packings leaks, the fluid seal can be restored by retightening the bolts. Retightening must be carried out while the valve is closed to prevent an unrestrained expansion of the seat packing into the valve bore.

The valve shown in Figure 3-20 differs from the one in Figure 3-19 only in that the piston is pressure balanced. The two packings around the piston are both seat packings, and a separate packing is provided for the stem. The

Figure 3-19. Piston valve, standard pattern, seat packing mounted in valve body, piston pressure unbalanced. (Courtesy of Rich. Klinger AG.)

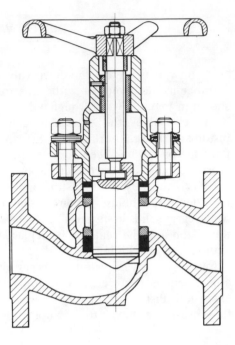

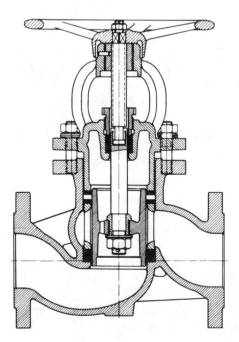

Figure 3-20. Piston valve, standard pattern, seat packing mounted in valve body, piston pressure balanced. (Courtesy of Rich. Klinger AG.)

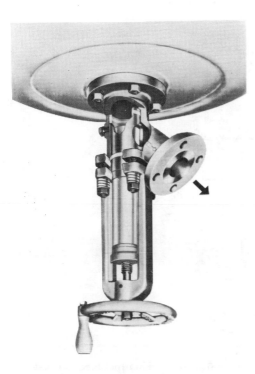

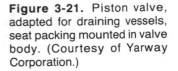

Figure 3-21. Piston valve, adapted for draining vessels, seat packing mounted in valve body. (Courtesy of Yarway Corporation.)

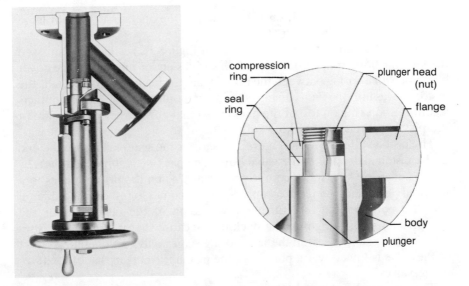

Figure 3-22. Piston valve adapted for draining vessels, seat packing mounted on piston; the "ram-seal" principle. (Courtesy of Fetterolf Corporation.)

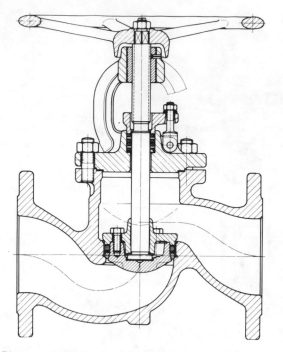

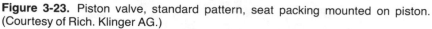

Figure 3-23. Piston valve, standard pattern, seat packing mounted on piston. (Courtesy of Rich. Klinger AG.)

purpose of balancing the piston is to minimize the operating effort in large valves operating against high fluid pressures.

The packing train of the valve shown in Figure 3-21 is likewise stressed through the bonnet in conjunction with springs under the nuts of the cover bolts, or with a spring between the bonnet and the packing. However, as the piston moves into the final closing position, a shoulder on the piston contacts a compression ring on top of the packing so that any further progression of the piston tightens the packing still further.

The piston valve shown in Figure 3-22 carries the seat packing on the end of the piston instead of in the valve bore. The packing is supported thereby on its underside by a loose compression ring. When the piston moves into the final closing position, the compression ring comes to rest on a shoulder in the seat bore so that any further progression of the piston causes the compression ring to tighten the packing. Because the packing establishes interference with the seat in the last closing stages only, the operating effort of the valve is lower over a portion of the piston travel than that of the foregoing valves.

The piston valve shown in Figure 3-23 also carries the seat packing on the piston. However, the loose compression ring is replaced by a friction ring which acts as a spring element and, as such, prestresses the packing. When the piston moves into the seat, the friction ring comes to rest in the seat bore, and any progression of the piston increases the packing stress.

Standards Pertaining to Piston Valves

Refer to Appendix C regarding standards pertaining to piston valves.

Applications

Duty:
  Controlling flow.
  Stopping and starting flow.
Service:
  Gases.
  Liquids.
  Fluids with solids in suspension.
  Vacuum.

PARALLEL GATE VALVES

Parallel gate valves are slide valves with a parallel-faced gate-like closure member. This closure member may consist of a single disc or twin discs with a spreading mechanism in-between. Typical valves of this type are shown in Figures 3-24 through 3-31.

The force which presses the disc against the seat is controlled by the fluid pressure acting on either a floating disc or a floating seat. In the case of twin-disc parallel gate valves, this force may be supplemented with a mechanical force from a spreading mechanism between the discs.

One advantage of parallel gate valves is their low resistance to flow, which in the case of full-bore valves approaches that of a short length of straight pipe. Because the disc slides across the seat face, parallel gate valves are also capable of handling fluids which carry solids in suspension.

This mode of valve operation also imposes some limitations to the use of parallel gate valves:

- If fluid pressure is low, the seating force may be insufficient to produce a satisfactory seal between metal-to-metal seatings.
- Frequent valve operation may lead to excessive wear of the seating faces, depending on magnitude of fluid pressure, width of seating faces, lubricity of the fluid to be sealed, and the wear resistance of the seating

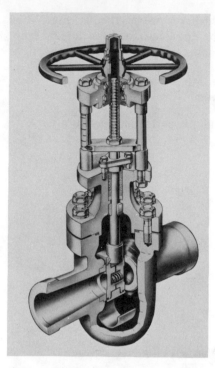

Figure 3-24. Parallel slide gate valve with converging-diverging flow passage and follower eye piece. (Courtesy of Hopkinsons Limited.)

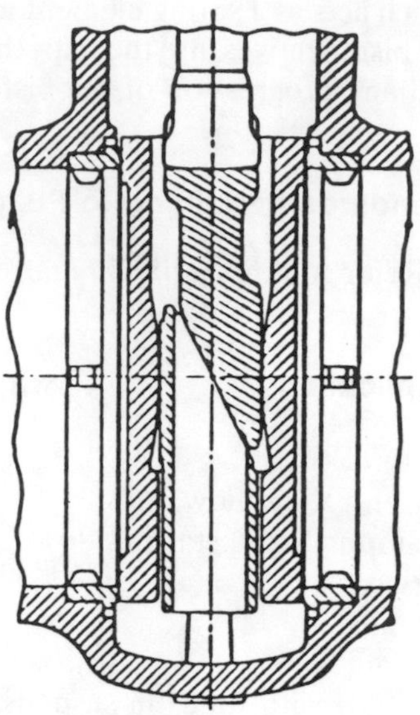

Figure 3-25. Scrap view of parallel gate valve showing double-disc closure member with wedging mechanism. (Courtesy of Pacific Valves, Inc.)

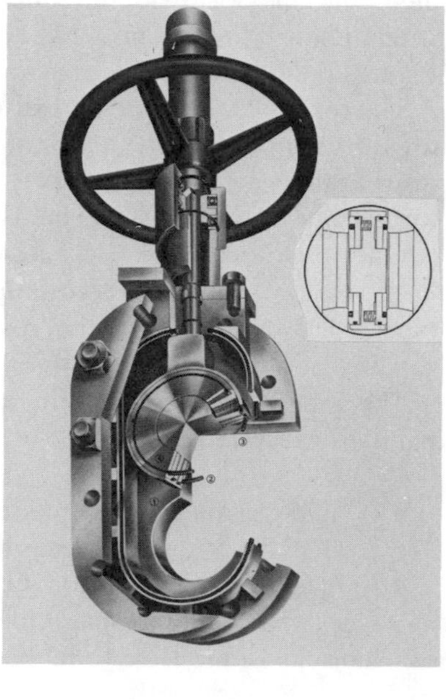

Figure 3-26. Parallel gate valve with scrap view of seating arrangement showing spring-loaded floating inserts in disc. (Courtesy of Grove Valve and Regulator Company.)

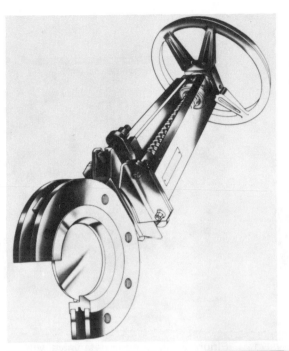

Figure 3-27. Knife gate valve.
(Courtesy of DeZurik.)

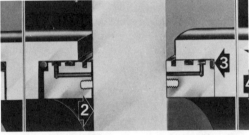

Figure 3-28. Conduit gate valve with scrap view
of seating arrangement showing floating seats.
(Courtesy of W.K.M. Valve Division, ACF Industries, Inc.)

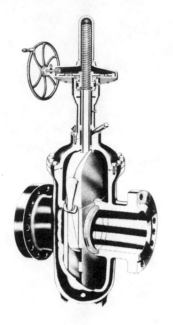

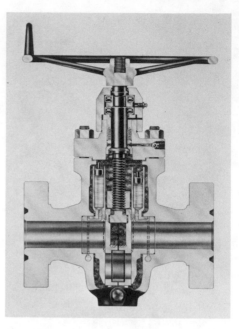

Figure 3-29. Conduit gate valve with floating seats and expandable disc. (Courtesy of W.K.M. Valve Division, ACF Industries, Inc.)

Figure 3-30. Conduit gate valve with automatic injection of a sealant to the downstream seatings each time the valve is closing. (Courtesy of McEvoy Oilfield Equipment Company.)

material. For this reason, parallel gate valves are normally used for infrequent valve operation only.

- Loosely guided discs and loose disc components will tend to rattle violently when shearing high density and high velocity flow.
- Flow control from a circular disc traveling across a circular flow passage becomes satisfactory only between the 50% closed and the fully closed positions. For this reason, parallel gate valves are normally used for on-off duty only, though some types of parallel gate valves have also been adopted for flow control, for example by V-porting the seat.

The parallel gate valves shown in Figures 3-24 through 3-27 are referred to as conventional parallel gate valves, and those of Figures 3-28 through 3-31 are referred to as conduit gate valves. The latter are full-bore valves which differ from the former in that the disc seals the valve body cavity against the ingress of solids in both the open and closed valve positions. Such valves may therefore be used in pipelines which have to be scraped.

Conventional Parallel Gate Valves

The valves shown in Figures 3-24 through 3-27 are representative of the common varieties of conventional parallel gate valves.

One of the best known is the valve shown in Figure 3-24, commonly referred to as a parallel slide gate valve. The closure member consists of two discs with springs in-between. The duties of these springs are to keep the upstream and downstream seatings in sliding contact and to improve the seating load at low fluid pressures. The discs are carried in a belt eye in a manner which prevents their unrestrained spreading as they move into the fully open valve position.

The flow passage of this particular parallel slide gate valve is venturi shaped. The gap between the seats of the fully open valve is bridged by an eyelet to ensure a smooth flow through the valve. The advantages offered by this construction include not only economy of construction but also a reduced operating effort and lower maintenance cost. The only disadvantage is a slight increase in pressure loss across the valve.

The seating stress reaches its maximum value when the valve is nearly closed, at which position the pressure drop across the valve is near maximum; but the seating area in mutual contact is only a portion of the total seating area. As the disc travels between the three-quarter closed to the

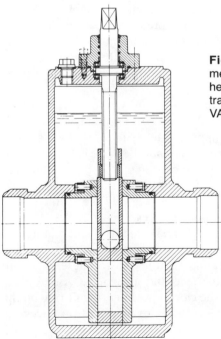

Figure 3-31. Conduit gate valve with metal seatings and oil-filled body cavity for heavily dust-laden gases and the hydraulic transport of coal and ores. (Courtesy of VAG-Armaturen GmbH.)

nearly closed valve position, the flowing fluid tends to tilt the disc into the seat bore, so heavy wear may occur in the seat bore and on the outer edge of the disc. To keep the seating stress and corresponding seating wear within acceptable limits, the width of the seatings must be made appropriately wide. Although this requirement is paradoxical in that the seating width must be small enough to achieve a seat seal but wide enough to keep seating wear within acceptable limits, the fluid tightness which is achieved by these valves satisfies the leakage criterion of the steam class, provided the fluid pressure is not too low.

Parallel slide gate valves have excellent advantages on other points: the seatings are virtually self aligning and the seat seal is not impaired by thermal movements of the valve body. Also, when the valve has been closed in the cold condition, thermal extension of the stem cannot overload the seatings. Furthermore, when the valve is being closed, a high accuracy in the positioning of the discs is not necessary, thus an electric drive for the valve can be travel limited. Because an electric drive of this type is both economical and reliable, parallel slide gate valves are often preferred as block valves in larger power stations for this reason alone. Of course, parallel slide gate valves may be used also for many other services such as water — in particular boiler feed water — and oil.

A variation of the parallel slide gate valve used mainly in the USA is fitted with a closure member like the one shown in Figure 3-25. The closure member consists of two discs with a wedging mechanism in-between, which, on contact with the bottom of the valve body, spreads the discs apart. When the valve is being opened again, the wedging mechanism releases the discs. Because the angle of the wedge must be wide enough for the wedge to be self releasing, the supplementary seating load from the wedging action is limited.

To prevent the discs from spreading prematurely, the valve must be mounted with the stem upright. If the valve must be mounted with the stem vertically down, the wedge must be appropriately supported by a spring.

The performance characteristic attributed to parallel slide gate valves also applies largely to this valve. However, solids carried by the flowing fluid and sticky substances may interfere with the functioning of the wedging mechanism. Also, thermal extension of the stem can overload the seatings. The valve is used mainly in gas, water, and oil services.

Conventional parallel gate valves may also be fitted with soft seatings, as in the valve shown in Figure 3-26. The closure member consists here of a disc which carries two spring-loaded floating seating rings. These rings are provided with a bonded O-ring on the face and a second O-ring on the periphery. When the disc moves into the closed position, the O-ring on the face of the floating seating ring contacts the body seat and produces the initial fluid seal. The fluid pressure acting on the back of the seating ring then forces the seatings into still closer contact.

Because the unbalanced area on the back of the floating rings is smaller than the area of the seat bore, the seating load for a given fluid pressure and valve size is smaller than in the previously described valves. However, the valve achieves a high degree of fluid tightness by means of the O-ring even at low fluid pressures. This sealing principle also permits double block and bleed.

The parallel gate valve shown in Figure 3-27 is known as the knife gate valve, and is designed to handle slurries, including fibrous material. The valve owes its ability to handle these fluids to the knife-edged disc which is capable of cutting through fibrous material, and the virtual absence of a valve body cavity. The disc travels in lateral guides and is forced against the seat by lugs at the bottom. If a high degree of fluid tightness is required, the valve may also be provided with an O-ring seat seal.

Conduit Gate Valves

Figures 3-28 through 3-31 show four types of valves which are representative of conduit gate valves. All four types of valves are provided with floating seats which are forced against the disc by the fluid pressure.

The seats of the conduit gate valve shown in Figure 3-28 are faced with PTFE and sealed peripherally by O-rings. The disc is extended at the bottom to receive a port hole. When the valve is fully open, the port hole in the disc engages the valve ports so that the disc seals the valve body cavity against the ingress of solids. The sealing action of the floating seats also permits double block and bleed. If the seat seal should fail in service, a temporary seat seal can be produced by injecting a sealant to the seat face.

The conduit gate valve shown in Figure 3-29 differs from the previous one in that the disc consists of two halves with a wedge-shaped interface. These halves are interlinked so that they wedge apart when being moved into the fully open or closed positions, but relax in the intermediate position to permit the disc to travel. Depending on the use of the valve, the face of the floating seats may be metallic or provided with a PTFE insert. To prevent the ingress of solids into the valve body cavity during all stages of disc travel, the floating seats are provided with skirts, in-between which the disc travels. This valve likewise permits double block and bleed. Also, should the seat seal fail in service, a temporary seat seal can be provided by injecting a sealant to the seat face.

The sealing action of the conduit gate valve shown in Figure 3-30 depends on a sealant which is fed to the downstream seat face each time the valve is operated. For this purpose, the floating seats carry reservoirs which are filled with a sealant and topped by a floating piston. The entire valve chamber is, furthermore, filled with a grease which transmits the fluid pressure to the top of the piston of the downstream reservoir. The closure member consists of two discs which are spread apart by springs. Both the sealant and the body grease can be replenished from the outside while the valve is in service. Each reservoir filling is sufficient for over 100 valve operations. This mode of sealing achieves a high degree of fluid tightness at high fluid pressures.

The conduit gate valve shown in Figure 3-31 is especially designed for the hydraulic transport of coal and ore, and the transport of heavily dust-laden gases. The gate, which consists of a heavy wear-resistant plate with a port hole in the bottom, slides between two floating seats which are highly prestressed against the disc by means of disc springs. To prevent any possible entry of solids into the body cavity, the seats are provided with skirts for the full travel of the disc. The faces of the disc and seats are metallic and highly polished. The valve chamber is, furthermore, filled with a lubricant which ensures lubrication of the seating faces.

Valve Bypass

The seating load of the larger parallel gate valves (except those with floating seats) can become so high at high fluid pressures that friction between the seatings can make it difficult to raise the disc from the closed position. Such valves are therefore frequently provided with a valved bypass line which is used to relieve the seating load prior to opening the valve. There are no fast rules about when to employ a bypass, and the manufacturer's recommendation may be sought. Some standards of gate valves contain recommendations on the minimum size of the bypass.

In the case of gases and vapors, such as steam, which condense in the cold downstream system, the pressurization of the downstream system can be considerably retarded. In this instance, the size of the bypass line should be larger than the minimum recommended size.

Pressure-Equalizing Connection

In the case of the conventional double-seated parallel gate valves shown in Figures 3-24 and 3-25, thermal expansion of a liquid trapped in the closed valve chamber will force the upstream and downstream discs into more intimate contact with their seats, and cause the pressure in the valve chamber to rise. The higher seating stress makes it in turn more difficult to raise the discs, and the pressure in the valve chamber may quickly become high enough to cause a bonnet flange joint to leak or the valve body to deform.

Thus, if such valves are used to handle a liquid with high thermal expansion, they must be provided with a pressure-equalizing connection, which connects the valve chamber with the upstream piping.

The pressure rise in the valve chamber may also be caused by the revaporation of trapped condensate, as in the case where these valves are closed against steam. Both the valve chamber and the upstream piping are initially under pressure and filled with steam. Eventually, the steam will cool, condense, and be replaced to some extent with air.

Upon restart, the steam will enter the upstream piping and — since the upstream seat is not normally fluid tight against the upstream pressure — will enter the valve chamber. Some of the new steam will also condense initially until the valve body and the upstream piping have reached the saturation temperature of the steam.

When this has happened, the steam begins to boil off the condensate. If no pressure-equalizing connection is provided, the expanding steam will force the upstream and downstream discs into more intimate contact with their seats, and raise the pressure in the valve chamber. The magnitude of the developing pressure is a function of the water temperature and the degree of filling of the valve chamber with water, and may be obtained from Figure 3-32.

The pressure-equalizing connection may be provided by a hole in the upstream disc or by other internal or external means. Some makers of parallel gate valves of the types shown in Figures 3-24 and 3-25 combine the bypass line with a pressure-equalizing line if the valve is intended for steam.

Standards Pertaining to Parallel Gate Valves

A list of USA and British standards pertaining to parallel gate valves may be found in Appendix C.

Applications

Duty:
    Stopping and starting flow.
    Infrequent operation.
Service:
    Gases.
    Liquids.
    Fluids with solids in suspension.
    Knife gate valve for slurries, fibers, powders, and granules.
    Vacuum.
    Cryogenic.

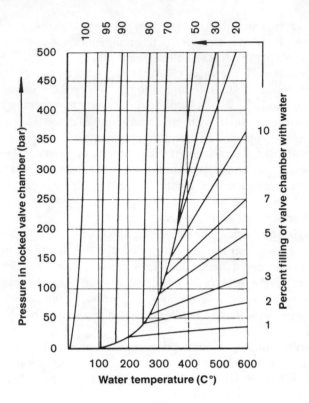

Figure 3-32. Pressure in locked valve chamber as a result of the revaporation of trapped water condensate. (Courtesy of Sempell Armaturen)

WEDGE GATE VALVES

Wedge gate valves differ from parallel gate valves in that the closure member is wedge-shaped instead of parallel, as shown in Figures 3-33 through 3-43. The purpose of the wedge shape is to introduce a high supplementary seating load which enables metal-seated wedge gate valves to seal not only against high but also low fluid pressures. The degree of seat tightness which can be achieved with metal-seated wedge gate valves is therefore potentially higher than with conventional metal-seated parallel gate valves. However, the upstream seating load due to wedging is not normally high enough to achieve an upstream seat seal with metal-seated wedge gate valves.

The bodies of these valves carry guide ribs, or slots, in which the disc travels. The main purpose of these guides is to carry the wedge away from the

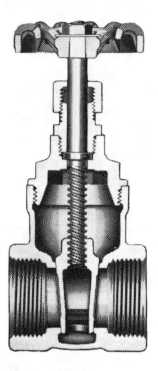

Figure 3-33. Wedge gate valve with plain hollow wedge, screwed-in bonnet, internal screw. (Courtesy of Crane Co.)

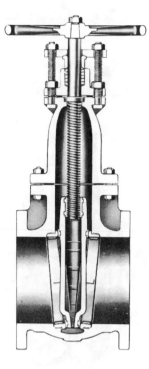

Figure 3-34. Wedge gate valve with plain hollow wedge, bolted bonnet, internal screw. (Courtesy of Crane Co.)

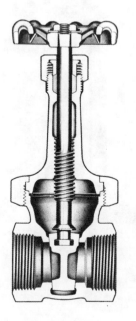

Figure 3-35. Wedge gate valve with plain solid wedge, union bonnet, internal screw. (Courtesy of Crane Co.)

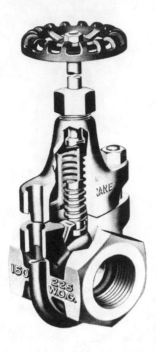

Figure 3-36. Wedge gate valve with clamped bonnet, internal screw. (Courtesy of Crane Co.)

Figure 3-37. Wedge gate with plain solid wedge, bolted bonnet, external screw. (Courtesy of Crane Co.)

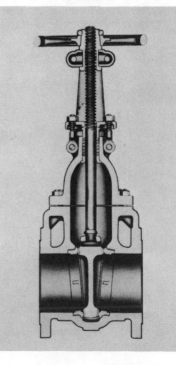

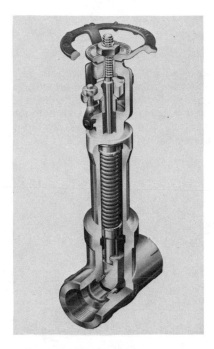

Figure 3-39. Seat of wedge gate valve with PTFE sealing insert. (Courtesy of Crane Co.)

Figure 3-38. Wedge gate with plain solid wedge, welded bonnet, external screw, bellows stem seal, and auxiliary stuffing-box seal. (Courtesy of Pegler Hattersley Limited.)

Figure 3-40. Wedge gate valve with flexible wedge, pressure-sealed bonnet, external screw. (Courtesy of Crane Co.)

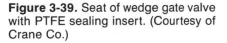
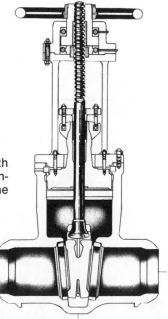

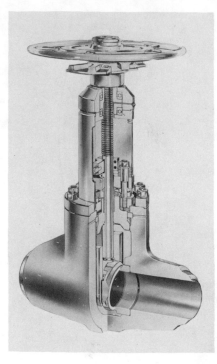

Figure 3-40a. Forged wedge gate valve with pressure sealed bonnet, incorporating flexible wedge with hard faced grooves sliding on machined body ribs. (Courtesy of Veelan Engineering Limited.)

Figure 3-41. Wedge gate valve with two-piece wedge, pressure-sealed bonnet, external screw. (Courtesy of Flow Control Division, Rockwell International.)

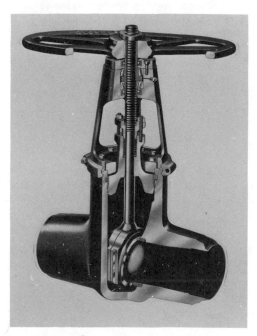

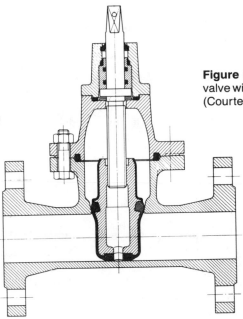

Figure 3-42. Rubber-seated wedge gate valve without cavity in bottom of valve body. (Courtesy of VAG-Armaturen GmbH.)

Figure 3-43. Rubber-seated wedge gate valve without cavity in bottom of valve body. (Courtesy of Schmitz & Schulte, D-5093 Burscheid.)

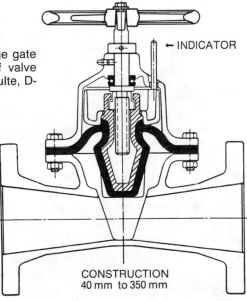

← INDICATOR

CONSTRUCTION
40 mm to 350 mm

downstream seat except for some distance near the closed valve position so as to minimize wear between the seatings. A second purpose of the guides is to prevent the disc from rotating excessively while traveling between the open and closed valve positions. If some rotation occurs, the disc will initially jam on one side between the seats and then rotate into the correct position before traveling into the final seating position.

There are also types of wedge gate valves that can dispense with a wedge guide, such as the valve shown in Figure 3-43 in which the wedge is carried by the diaphragm.

Compared with parallel gate valves, wedge gate valves also have some negative points:

- Wedge gate valves cannot accommodate a follower conduit as conveniently as parallel gate valves can.
- As the disc approaches the valve seat, there is some possibility of the seatings trapping solids carried by the fluid. However, rubber-seated wedge gate valves as shown in Figures 3-42 and 3-43 are capable of sealing around small trapped solids.
- An electrical drive for wedge gate valves is more complicated than for parallel gate valves in that the drive must be torque limited instead of travel limited. The operating torque of the drive must thereby be high enough to effect the wedging of the wedge into the seats while the valve is being closed against the full differential line pressure. If the valve is closed against zero differential pressure, the wedging of the wedge into the seats becomes accordingly higher. To permit the valve to be opened again against the full differential pressure, and to allow also for a possible increase of the operating effort due to thermal movements of the valve parts, the operator must be generously sized.

The limitations of wedge gate valves are otherwise similar to those of parallel gate valves.

Efforts to improve the performance of wedge gate valves led to the development of a variety of wedge designs; the most common ones are described in the following section.

Variations of Wedge Design

The basic type of wedge is the plain wedge of solid or hollow construction, as in the valves shown in Figures 3-33 through 3-38. This design has the advantage of being simple and robust, but distortions of the valve body due to thermal and pipeline stresses may unseat or jam the metal-seated wedge. A failure of this kind is more often experienced in valves of light-weight construction.

The sealing reliability of gate valves with a plain wedge can be improved by elastomeric or plastic sealing elements in either the seat or the wedge. Figure 3-39 shows a seat in which the sealing element is a PTFE insert. The PTFE insert stands proud of the metal face just enough to ensure a seal against the wedge.

Efforts to overcome the alignment problem of plain wedges led to the development of self-aligning wedges; Figures 3-40, 3-40a, and 3-41 show typical examples. The simplest of these is the flexible wedge shown in Figures 3-40 and 3-40a which is composed of two discs with an integral boss in-between. The wedge is sufficiently flexible to find its own orientation. Because the wedge is simple and contains no separate components which could rattle loose in service, this construction has become a favored design.

The self-aligning wedge of the valve shown in Figure 3-41 consists of two identical tapered plates which rock around a separate spacer ring. This spacer ring may also be used to adjust the wedge assembly for wear. To keep the plates together, the body is provided with grooves in which the wedge assembly travels.

Rubber lining of the wedge, as in the valves shown in Figures 3-42 and 3-43 led to the development of new seating concepts in which the seat seal is achieved in part between the rim of the wedge and the valve body. In this way it became possible to avoid altogether the creation of a pocket at the bottom of the valve body. These valves are therefore capable of handling fluids carrying solids in suspension which would otherwise collect in an open body cavity.

In the case of the valve shown in Figure 3-42, the wedge is provided with two stirrup-shaped rubber rings which face the rim of the wedge at the bottom, sides, and the top lateral faces. When the valve is being closed, the rubber rings seal against the bottom and the side walls of the valve body and, by a wedging action, against the seat faces at the top.

The wedge of the valve shown in Figure 3-43 is completely rubber lined and forms part of a diaphragm which separates the operating mechanism from the flowing fluid. When the valve is being closed, the bottom of the wedge seals against the bottom of the valve body, while the side and top portions of the wedge seal axially against the body seats. The valve may also be lined with corrosion-resistant materials, and is therefore widely used in the chemical industry.

Connection of Wedge to Stem

The wedge-to-stem connection usually consists of a T-slot in the top of the wedge which receives a collar on the stem. According to API standard 600, this connection must be stronger than the weakest stem section so that the wedge cannot become detached from the steam while operating the valve.

The T-slot in the wedge may thereby be oriented in line with the flow passage, as in the valves shown in Figures 3-37, 3-38, and 3-40, or across the flow passage as shown in Figures 3-35 and 3-40a. The latter construction permits a more compact valve body design and, therefore, has become popular for economic reasons. Also, this construction lowers favorably the point at which the stem acts on the wedge. However, the T-slot must be wide enough to accommodate the play of the wedge in its guide, allowing also for wear of the guide.

There are also exceptions to this mode of wedge-to-stem connection, as in the valves shown in Figures 3-33 through 3-36, where the stem must carry the entire thrust on the wedge. For this reason, this construction is suitable for low pressure applications only. In the valve shown in Figure 3-41, guide play is virtually absent, allowing the stem to be captured in the wedge.

Wedge Guide Design

The body guides commonly consist either of ribs which fit into slots of the wedge, or of slots which receive ribs of the wedge. Figures 3-40a and 3-41 illustrate these guiding mechanisms.

The body ribs are normally as cast for reason of low cost construction. However, the rough surface finish of such guides is not suited for carrying the traveling wedge under high load. For this reason, the wedge is carried on valve opening initially on the seat until the fluid load has become small enough for the body ribs to carry the wedge. This method of guiding the wedge may require considerable play in the guides which must be matched by the play in the T-slot for the wedge suspension.

Once the body ribs begin to carry the wedge upon valve opening, the wedge must be fully supported by the ribs. If the length of support is insufficient, the force of the flowing fluid acting on the unsupported section of the wedge may be able to tilt the wedge into the downstream seat bore. This requirement is sometimes not complied with. On the other hand, some valve makers go to any length to ensure full wedge support.

There is no assurance that the wedge will slide on the stem collar when opening the valve. At this stage of valve operation, there is considerable friction between the contact faces of the T-slot and stem collar, possibly causing the wedge to tilt on the stem as the valve opens. If, in addition, the fit between T-slot and stem collar is tight, and the fluid load on the disc is high, the claws forming the T-slot may crack.

For critical applications, guides in wedge-gate valves are machined to close tolerances and designed to carry the wedge over nearly the entire valve travel, as in the valves shown in Figures 3-40a and 3-41.

In the valve shown in Figure 3-40a, the wedge grooves are hard faced and precision guided on machined guide ribs welded to the valve body. The wedge is permitted in this particular design to be carried by the seat for 5% of the total travel.

In the valve shown in Figure 3-41, the wedge consists of two separate wedge-shaped plates. These carry hard-faced tongues that are guided in machined grooves of the valve body. When wear has taken place in the guides, the original guide tolerance can be restored by adjusting the thickness of a spacer ring between the two wedge plates.

Valve Bypass

Wedge gate valves may have to be provided with bypass connections for the same reason described for parallel gate valves on page 74.

Pressure-Equalizing Connection

In the case of wedge gate valves with a self-aligning double-seated wedge, thermal expansion of a fluid locked in the valve body will force the upstream and downstream seatings into still closer contact and cause the pressure in the valve body cavity to rise. A similar situation may arise with soft-seated wedge gate valves where the wedge is capable of producing an upstream seat seal. Thus, if such valves handle a liquid with high thermal expansion, or if revaporation of trapped condensate can occur, they may have to be provided with a pressure-equalizing connection as described for parallel gate valves on page 71.

Case Study of Wedge Gate Valve Failure

Figures 3-43a through 3-43c show components of a DN 300 (NPS 12) class 150 wedge gate valve to API standard 600 that failed on first application.

The valve was mounted in this case under an angle of 45° from the vertical for ease of hand operating a gear drive. This operating position required the wedge to ride on the body rib. Unfortunately, the guide slot in the wedge had sharp edges. As the wedge traveled on the rib, the sharp edges of the guide slot caught on the rough rib surface, causing the wedge to tilt until contacting the opposite body rib. At this stage, the wedge was jammed. Further closing effort by the valve operator produced the damage to the valve internals shown in Figures 3-43a through 3-43c.

Figure 3-43a shows the damage to the wedge guide that was riding on the body rib and Figure 3-43b the damaged body rib. Figure 3-43c shows the bent valve stem and damage to the stem surface around the stem guide bush.

Further inspection of the valve showed also that play in the wedge guides was larger than the possible travel of the wedge on the stem. Thus, the stem

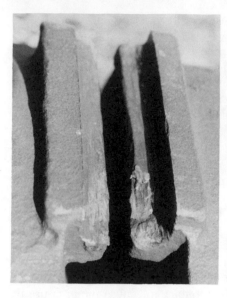

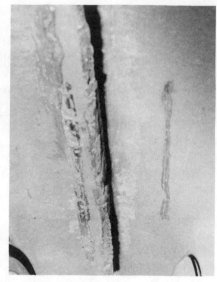

Figure 3-43a. Damage to guide slots of flexible wedge, resulting from attempted closure of wedge gate valve with wedge rotated and jammed in valve body, size DN 300 (NPS 12) class 150.

Figure 3-43b. Damage to valve body guide ribs, resulting from attempted closure of wedge gate valve with rotated wedge jammed in valve body, size DN 300 (NPS 12) class 150.

Figure 3-43c. Bent valve stem with damage to stem surface, resulting from attempted closure of wedge gate valve with rotated wedge jammed in valve body, size DN 300 (NPS 12) class 150.

had to carry the wedge for part of its travel in a tilted position. Furthermore, the lengths of body rib and wedge guide were far too short to adequately support the wedge during all stages of travel.

This valve failure was not isolated but was typical for a high percentage of all installed wedge gate valves. Finally, all suspect valves had to be replaced prior to start-up of the plant.

Standards Pertaining to Wedge Gate Valves

A list of USA and British standards pertaining to wedge gate valves may be found in Appendix C.

Applications

Duty:
    Stopping and starting flow.
    Infrequent operation.
Service:
    Gases.
    Liquids.
    Rubber-seated wedge gate valves without bottom cavity for fluids carrying solids in suspension.
    Vacuum.
    Cryogenic.

PLUG VALVES

Plug valves are rotary valves in which a plug-shaped closure member is rotated through increments of 90° to engage or disengage a port hole or holes in the plug with the ports in the valve body. The shape of the plug may thereby by cylindrical, as in the valves shown in Figures 3-44 through 3-48, or tapered, as in the valves shown in Figures 3-49 through 3-53. Rotary valves with a ball-shaped plug are likewise a members of the plug valve family, but are conventionally referred to as ball valves. These valves are discussed separately on page 98.

The shape of the port is commonly rectangular in parallel plugs and truncated triangular in taper plugs. These shapes permit a slimmer valve construction of reduced weight, but at the expense of some pressure drop. Full area round-bore ports are normally used only if the pipeline has to be scraped or the nature of the fluid demands a full area round bore. However, some plug valves are made only with round-bore because of the method of sealing employed.

Figure 3-44. Lubricated cylindrical plug valve. (Courtesy of Pegler Hattersley Limited.)

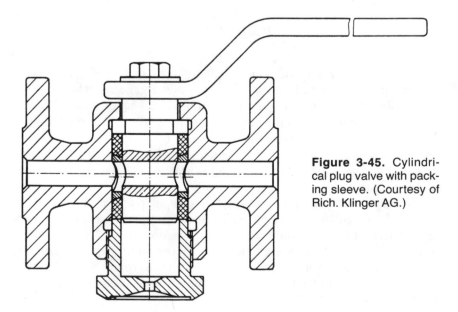

Figure 3-45. Cylindrical plug valve with packing sleeve. (Courtesy of Rich. Klinger AG.)

Figure 3-46. Cylindrical plug valve with expandable split plug. (Courtesy of Langley Alloys Limited.)

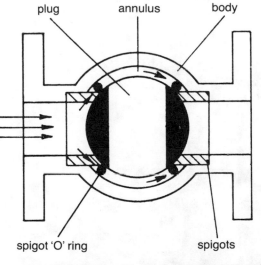

VALVE SHOWN IN CLOSED POSITION

Figure 3-47. Cylindrical plug valve with O-ring seat seal. Valve shown in closed position. (Courtesy of Orseal Limited.)

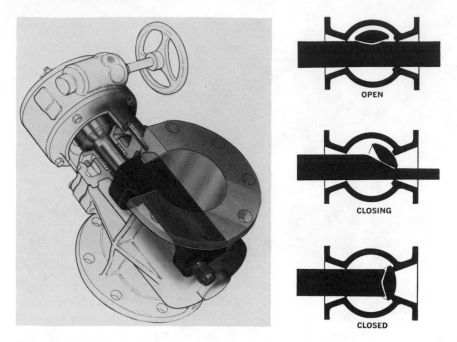

Figure 3-48. Cylindrical plug valve with eccentric semi-plug. (Courtesy of De-Zurik.)

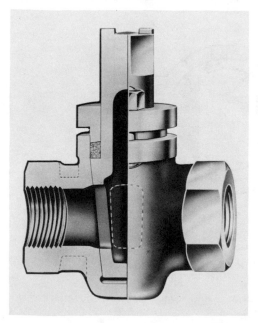

Figure 3-49. Taper plug valve with unlubricated metal seatings. (Courtesy of Pegler Hattersley Limited.)

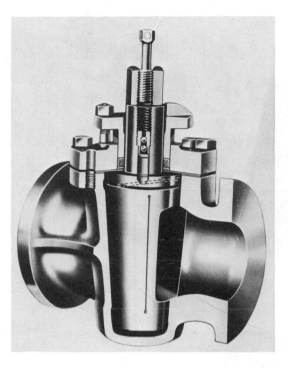

Figure 3-50. Lubricated taper plug valve. (Courtesy of Serck Audco Pty. Ltd.)

Figure 3-51. Lubricated taper plug valve with inverted pressure-balanced plug. (Courtesy of Flow Control Division, Rockwell International.)

Figure 3-52. Taper plug valve with PTFE body sleeve. (Courtesy of Xomox Corporation, "Tufline.")

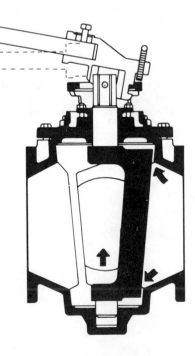

Figure 3-53. Multiport taper plug valve with lift plug. (Courtesy of DeZurik.)

Plug valves are best suited for stopping and starting flow and flow diversion, though they are also used occasionally for moderate throttling, depending on the nature of the service and the erosion resistance of the seatings. Because the seatings move against each other with a wiping motion, and in the fully open position are also fully protected from contact with the flowing fluid, plug valves are generally capable of handling fluids with solids in suspension.

Cylindrical Plug Valves

The use to which plug valves can be put depends to some extent on the way the seal between the plug and the valve body is produced. In the case of cylindrical plug valves, four sealing methods are frequently employed: by a sealing compound, by expanding the plug, by O-rings, and by wedging an eccentrically shaped plug into the seat.

The cylindrical plug valve shown in Figure 3-44 is a lubricated plug valve in which the seat seal depends on the presence of a sealing compound between the plug and the valve body. The sealing compound is introduced to the seatings through the shank of the plug by a screw or an injection gun. Thus, it is possible to restore a defective seat seal by injecting an additional amount of sealing compound while the valve is in service.

Because the seating surfaces are protected in the fully open position from contact with the flowing fluid, and a damaged seat seal can easily be restored, lubricated plug valves have been found to be particularly suitable for abrasive fluids. However, lubricated plug valves are not intended for throttling, although they are sometimes used for this purpose. Because throttling removes the sealing compound from the exposed seating surfaces, the seat seal must be restored, in this case, each time the valve is closed.

Unfortunately, the manual maintenance of the sealing compound is often a human problem. Automatic injection can overcome this problem, but it adds to the cost of installation. When the plug has become immovable in the valve body due to lack of maintenance or improper selection of the sealing compound, or because crystallization has occurred between the seatings, the valve must be cleaned or repaired.

The seat seal of the cylindrical plug valves shown in Figures 3-45 and 3-46 depends on the ability of the plug to expand against the seat.

The plug of the valve shown in Figure 3-45 is fitted for this purpose with a packing sleeve which is tightened against the seat by a follower nut. The packing commonly consists of compressed asbestos or solid PTFE. If the packing needs retightening to restore the seat tightness, this must be carried out while the valve is in the closed position to prevent the packing from expanding into the flow passage. The valve is made only in small sizes, but may be used for fairly high pressures and temperatures. Typical applications for the valve are the isolation of pressure gauges and level gauges.

The plug of the valve shown in Figure 3-46 is split into two halves and spread apart by a wedge which may be adjusted from the outside. The seat seal is provided here by narrow PTFE rings which are inserted into the faces of the plug halves. By this method of sealing and seat loading, the valve is also capable of double block and bleed.

The valve is specifically intended for duties for which stainless steel and other expensive alloys are required, and it is capable of handling slurries, but not abrasive solids. Very tacky substances also tend to render the wedging mechanism inoperable. But within these limits, the valve has proved to be very reliable under conditions of frequent valve operation.

Figure 3-47 shows a parallel plug valve in which the seat seal is provided by O-rings. The O-rings are mounted on spigot-like projections of the upstream and downstream ports which form the seats for the plug. When the valve is being closed, the fluid pressure enters the cavity between the plug and the valve body past the upstream O-ring and forces the downstream O-ring into intimate contact with the plug and the projections of the valve ports. The main application of this valve is for high-pressure hydraulic systems.

Efforts to eliminate most of the friction between the seatings and to control the sealing capacity of the valve by the applied torque led to the development of the eccentric plug valve shown in Figure 3-48, in which an eccentric plug matches a raised face in the valve body. By this method of seating, the valve is easy to operate and capable of handling sticky substances which are difficult to handle with other types of valves. The seating faces may be in metallic contact or lined with an elastomer or plastic in conjunction with the lining of the valve body.

Taper Plug Valves

Taper plug valves permit the leakage gap between the seatings to be adjusted by forcing the plug deeper into the seat. The plug may also be rotated while in intimate contact with the valve body, or lifted out of the valve body prior to rotation and seated again after being rotated through 90°.

Figure 3-49 shows a taper plug valve with unlubricated metal seatings. Because the friction between the seatings is high in this case, the permissible seating load is limited to keep the plug freely movable. The leakage gap between the seatings is therefore relatively wide, so the valve achieves a satisfactory seat seal only with liquids which have a high surface tension or viscosity. However, if the plug has been coated with a grease prior to installation, the valve may be used also for wet gases such as wet and oily compressed air.

The lubricated taper plug valve shown in Figure 3-50 is similar to the lubricated cylindrical plug valve except for the shape of the plug. Both valves may also serve the same duties. However, the lubricated taper plug valve has one operational advantage. Should the plug become immovable after a prolonged static period or as a result of neglected lubrication, the injection of additional sealing compound can lift the plug off the seat just enough to allow the plug to be moved again. When freeing the plug in this way, the gland should not be slackened as is sometimes done by users, but should rather be judiciously tightened. On the debit side, it is also possible to manually overtighten the plug and cause the plug to seize in this way.

The valve shown in Figure 3-51 is a lubricated taper plug valve in which the plug is mounted in the inverted position and divorced from the stem. The plug is adjusted in its position by a screw in the valve cover while the sealant is injected into the body at the stem end. To prevent the fluid pressure from driving the plug into the seat, the plug ends are provided with balance holes which permit the fluid pressure to enter the cavities at both plug ends. By this design the plug valve may be used for very high pressures without becoming inoperable because of the fluid pressure driving the plug into the seat.

Efforts to overcome the maintenance problem of lubricated plug valves led to the development of the taper plug valve shown in Figure 3-52 in which the plug moves in a PTFE body sleeve. The PTFE body sleeve keeps the plug from sticking, but the operating torque can still be relatively high due to the large seating area and the high seating stress. On the other hand, the large seating area gives good protection against leakage should some damage occur to the seating surface. As a result, the valve is rugged and tolerates abusive treatment. The PTFE sleeve also permits the valve to be made in exotic materials which would otherwise tend to bind in mutual contact. In addition, the valve is easily repaired in the field, and no lapping of the plug is required.

The taper plug valve shown in Figure 3-53 is designed to eliminate most of the sliding between the seatings. This is achieved by lifting the plug out of the body seat by means of a hinged lever arrangement prior to it being rotated, and reseating it after it is rotated to the desired position. The plug may be rubber faced, and is, by its taper, normally self locking. This particular valve is a multiport companion valve to the eccentric plug valve shown in Figure 3-48.

Antistatic Device

In plug valves, seats and packings made of a polymeric material such as PTFE can electrically insulate the plug and the valve stem from the valve body. Under these conditions, friction from the flowing fluid may generate an electrostatic charge in the plug and the stem which is high enough to produce an incendiary spark. This possibility is more pronounced with two-phase flow. If the fluid handled by the valve is flammable, the valve must be provided with an antistatic device which achieves electrical continuity between the plug, stem, and the valve body.

Plug Valves for Fire Exposure

Plug valves, which may be exposed to a plant fire when handling a flammable fluid, must remain essentially fluid-tight internally and externally and be operable during and after a fire. This degree of resistance to fire damage is particularly difficult to achieve where the plug valve normally relies on polymers for seat and stem seal. Common practice in this case is to provide the valve with an auxiliary metal seat, in close proximity to the plug, against which the plug can float after the soft seat has disintegrated. The soft stem packing can readily be replaced with a fire-resistant packing.

The requirements for testing and evaluating the performance of valves exposed to fire are similar to those for ball valves, described on page 105.

Multiport Configuration

Plug valves adapt readily to multiport configurations such as those shown in Figure 3-59. The valves may be designed for transflow, in which case the second flow passage opens before the first closes; or for non-transflow, in which case the first flow passage closes before the second opens. The transflow sequence is intended for duties in which the flow cannot be momentarily interrupted; for example, on the outlet of a positive displacement pump which is not protected by a relief valve. The non-transflow sequence may be required where a momentary flow into one port from the other is not permissible; for example, at the outlet of a measuring vessel; but the plug is not normally intended to shut-off fluid tight in the intermediate position. In most practical applications, however, where valves are operated fairly quickly, there is little difference between the effects of transflow and non-transflow in fluid flow.

The direction of flow through lubricated multiport valves should be such that the fluid pressure forces the plug against the port which is to be shut-off. If the pressure acts from the opposite direction, lubricated plug valves will not hold their rated pressure.

Face-to-Face Dimensions and Valve Patterns

The designers of the USA and British standards of cast iron and carbon steel plug valves have attempted to make the face-to-face dimensions of plug

positions of
three-way
L-ported
valves

positions of
three-way
T-ported
valves

positions of
four-way
valves

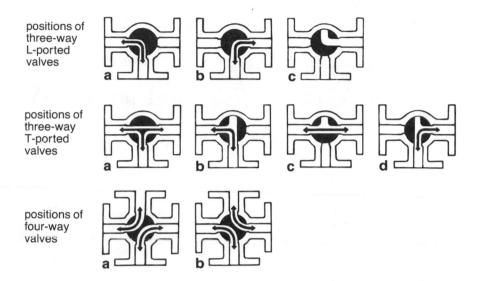

Figure 3-54. Multiport configurations of plug valves. (Courtesy of Serck Audco Pty. Ltd.)

valves identical to those of gate valves. To accommodate plug valves within these dimensions, some concessions had to be made on the flow area through the plug in the lower pressure ratings. But even with these concessions, plug valves for class 150 can be made interchangeable with gate valves up to DN 300 (NPS 12) only. This limitation led to the introduction of an additional long-series plug valve for class 150. As a result, the following valve patterns have emerged:

1. Short pattern, having reduced area plug ports and face-to-face dimensions that are interchangeable with gate valves. This pattern applies to sizes up to DN 300 (NPS 12) in class 150 only.
2. Regular pattern, having plug ports with areas larger than short or venturi patterns. The face-to-face dimensions are interchangeable with gate valves for pressure ratings class 300 and higher. Regular pattern plug valves for class 150 have face-to-face dimensions to a long series which are not interchangeable with gate valves.
3. Venturi pattern, having reduced area ports and body throats that are approximately venturi. The face-to-face dimensions are interchangeable with gate valves for pressure ratings class 300 and higher. Venturi-pattern plug valves for class 150 have face-to-face dimensions to a long series which are not interchangeable with gate valves.
4. Round-port full area pattern, having full area round ports through the valve. The face-to-face dimensions are longer than short, regular, or

venturi-pattern plug valves and are non-interchangeable with gate valves.

Standards Pertaining to Plug Valves

A list of USA and British standards pertaining to plug valves may be found in Appendix C.

Applications

Duty:
 Stopping and starting flow.
 Moderate throttling.
 Flow diversion.
Fluids:
 Gases.
 Liquids.
 Non-abrasive slurries.
 Abrasive slurries for lubricated plug valves.
 Sticky fluids for eccentric and lift plug valves.
 Sanitary handling of pharmaceuticals and food stuffs.
 Vacuum.

BALL VALVES

Ball valves are a species of plug valves having a ball-shaped closure member. The seat matching the ball is circular so that the seating stress is circumferentially uniform. Most ball valves are also equipped with soft seats which conform readily to the surface of the ball. Thus, from the point of sealing, the concept of the ball valve is excellent. The valves shown in Figures 3-55 through 3-61 are typical of the ball valves available.

The flow-control characteristic which arises from a round port moving across a circular seat and from the double pressure drop across the two seats is very good. However, if the valve is left partially open for an extended period under conditions of a high pressure drop across the ball, the soft seat will tend to flow around the edge of the ball orifice and possibly lock the ball in that position. Ball valves for manual control are therefore best suited for stopping and starting flow and moderate throttling. If flow control is automatic, the ball is continuously on the move, thus keeping this failure from normally occurring.

Because the ball moves across the seats with a wiping motion, ball valves will handle fluids with solids in suspension. However, abrasive solids will

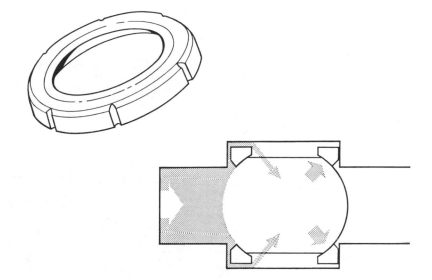

Figure 3-55. Schematic view of ball valve with floating ball and torsion seats, showing function of pressure-equalizing slots in periphery of seats. (Courtesy of Worchester Valve Co. Ltd.)

Figure 3-56. Ball valve with floating ball and torsion seats, with axial-entry body. (Courtesy of Jamesbury International Corp.)

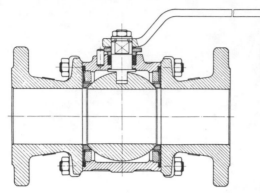

Figure 3-57. Ball valve with floating ball and diaphragm-supported seats, with sandwich-split body. (Courtesy of Rich. Klinger AG.)

Figure 3-58. Ball valve with wedge seats spring-loaded from the top, with one piece top-entry body. (Courtesy of Rockwell International.)

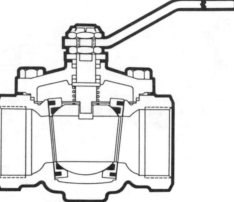

Figure 3-59. Ball valve with trunnion-supported ball and floating seats, with one-piece sealed body. (Courtesy of Cameron Iron Works, Inc.)

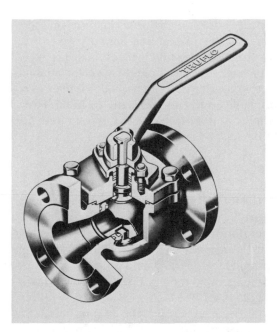

Figure 3-60. Ball valve with trunnion-supported ball and wedge seats forced into the valve body by the bonnet, with one-piece top-entry body. (Courtesy of Truflo Limited.)

Figure 3-61. Ball valve with cam mechanism to seat and unseat the ball. (Courtesy of Orbit Valve Company.)

damage the seats and the ball surface. Long, tough fibrous material may also present a problem, as the fibers tend to wrap around the ball.

To economize in the valve construction, most ball valves have a reduced bore with a venturi-shaped flow passage of about three-quarters the nominal valve size. The pressure drop across the reduced-bore ball valve is thereby so small that the cost of a full-bore ball valve is not normally justified. However, there are applications where a full-bore ball valve is required; for example, if the pipeline has to be scraped.

Seat Materials for Ball Valves

The most important seat material for ball valves is PTFE, which is inert to almost all chemicals. This property is combined with a low coefficient of friction, a wide range of temperature application, and excellent sealing properties. However, the physical properties of PTFE include also a high coefficient of expansion, susceptibility to cold flow, and poor heat transfer. The seat must therefore be designed around these properties. Plastic materials for ball valve seats also include filled PTFE, nylon, and many others. However, as the seating material becomes harder, the sealing reliability tends to suffer, particularly at low pressure differentials. Elastomers such as buna-N are also used for the seats, but they impose restrictions on fluid compatibility and range of temperature application. In addition, elastomers tend to grip the ball unless the fluid has sufficient lubricity. Metallic and carbon graphite seats are also used, but only to a small extent.

Seating Designs

The intimate contact between the seatings of ball valves may be achieved in a number of ways. Some of the ones more frequently used are:

1. By the fluid pressure forcing a floating ball against the seat, as in the valves shown in Figures 3-55 through 3-58.
2. By the fluid pressure forcing a floating seat ring against a trunnion-supported ball, as in the valve shown in Figure 3-59.
3. By relying mainly on the installed prestress between the seats and a trunnion-supported ball, as in the valve shown in Figure 3-60.
4. By means of a mechanical force which is introduced to the ball and seat on closing, as in the valve shown in Figure 3-61.

Ball valves are also available in which the seal between the seat and ball is achieved by means of a squeeze ring such as an O-ring.

The first sealing method, in which the seating load is regulated by the fluid pressure acting on the ball, is the most common one. The permissible operating pressure is limited in this case by the ability of the downstream seat ring to withstand the fluid loading at the operating temperature without permanent gross deformation.

The seat rings of the valves shown in Figures 3-55 and 3-56 are provided with a cantilevered lip which is designed so that the ball contacts initially only the tip of the lip. As the upstream and downstream seats are prestressed on assembly against the ball, the lips deflect and put the seat rings into torsion. When the valve is being closed against the line pressure, the lip of the downstream seat deflects still further until finally the entire seat surface matches the ball. By this design, the seats have some spring action which promotes good sealing action also at low fluid pressures. Furthermore, the resilient construction keeps the seats from being crushed at high fluid loads.

The seat rings of the valve shown in Figure 3-55 are provided with peripheral slots which are known as pressure-equalizing slots. These slots reduce the effect of the upstream pressure on the total valve torque. This is achieved by letting the upstream pressure filter by the upstream seat ring into the valve body cavity so that the upstream seat ring becomes pressure balanced.

The valves shown in Figures 3-57 and 3-58 are likewise designed to ensure a preload between the seatings. This is achieved in the valve shown in Figure 3-57 by supporting the seat rings on metal diaphragms which act as springs on the back of the seat rings. The seating preload of the valve shown in Figure 3-58 is maintained by a spring which forces the ball and seat-ring assembly from the top into wedge-shaped seat ring back faces in the valve body.

The design of the ball valve shown in Figure 3-59 is based on the second seating method in which the fluid pressure forces the seat ring against a trunnion-supported ball. The floating seat ring is sealed thereby peripherally by an O-ring. Because the pressure-uncompensated area of the seat ring can be kept small, the seating load for a given pressure rating can be regulated to suit the bearing capacity of the seat. These valves may therefore be used for high fluid pressures outside the range of floating-ball type ball valves. This particular valve is also provided with a device which rotates the seat rings by a small amount each time the valve is operated. The purpose of this rotating action is to evenly distribute the seat wear. Should the seat seal fail, a temporary seat seal can be provided by the injection of a sealant to the seatings.

The third seating method, in which the seat seal depends mainly on the installed prestress between the seats and a trunnion-supported ball, as in the valve shown in Figure 3-60 is designed to limit the operating torque of the valve at high fluid pressures. The lips around the ports of the ball are radiused (rounded) to reduce the seating interference when the ball is in the open position. When the ball is moved into the closed position, the seating interference increases. If the valve is required for double block and bleed, the back of the seat rings must be provided with an elastomeric O-ring.

The fourth seating method, in which the seating load is regulated on closing by an introduced mechanical force, is designed to avoid most of the sliding action between the seatings. In the valve shown in Figure 3-61, this is achieved by a cam mechanism which lifts the ball out of the seat prior to opening the valve and forces the ball back into the seat after closing the valve.

Pressure-Equalizing Connection

Double-seated ball valves may contain a sealed valve body cavity in both the open and closed valve positions. When the valve is closed, the sealed cavity extends between the upstream and downstream seats. When the valve is open, a sealed cavity may exist also between the ball and the valve body. If these cavities are filled with a liquid of high thermal expansion, the pressure rise in these cavities due to thermal expansion of the trapped fluid may overstress some valve components unless the excess fluid pressure can be relieved.

The cavity between the ball and the valve body is normally relieved to the flow passage via a hole in the top or bottom flank of the ball. If the valve is closed, the excess pressure in the cavity between the seats may be relieved in various ways.

In the case of ball valves with floating seats, as shown in Figure 3-59, excess pressure in the valve body will open the upstream seat seal where the least pressure differential exists. This permits the excess pressure to escape.

In other double-seated ball valves, however, the fluid pressure must overcome the prestress between the ball and the upstream seat. If the seat rings are provided with some springing action, as in the valves shown in Figures 3-55 through 3-57, the fluid pressure may be able to open the upstream seat seal without becoming excessively high. On the other hand, if the seat rings are of a more rigid construction, thermal expansion of the trapped fluid may create an excessively high pressure in the sealed cavity, depending on the prestress between the upstream seatings. In this case, the upstream flank of the ball is usually provided with a pressure-equalizing hole, thus permitting flow through the valve in one direction only. If the valve catalog does not advise on the need for a pressure-equalizing connection, the manufacturer should be consulted. The provision of a pressure-equalizing connection is not normally standard with ball valves except for cryogenic service. The pressure-equalizing connection is necessary in that case because of the rigidity of normally soft plastics at low temperatures which tends to resist the opening of the upstream seat seal.

Antistatic Device

The polymeric seats and packings used in ball valves can electrically insulate the ball and the stem from the valve body. Ball valves may therefore have to be provided with an antistatic device for the same reason as described on page 90 for plug valves. Figure 3-62 shows a typical antistatic device consisting of spring-loaded plungers — one fitted between the tongue of the stem and the ball, and a second between the stem and the body.

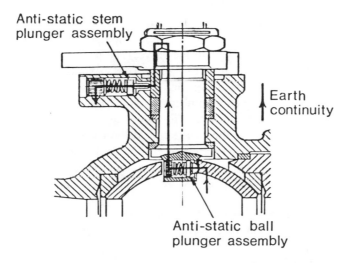

Figure 3-62. Antistatic device for grounding stem to ball and stem to body. (Courtesy of Worchester Valve Co., Ltd.)

Ball Valves for Fire Exposure

The soft seals for seat and stem commonly used in ball valves will perish if the valve is exposed to fire for a long enough period. If such valves are used for flammable fluids, they must be designed so that loss of the soft seals due to an external fire does not result in gross internal and external valve leakage. Such designs provide emergency seals for seat and stem that come into operation after the primary seals have failed.

The emergency seat-seal may be provided by a sharp edged or chamfered secondary metal seat in close proximity to the ball, so that the ball can float against the metal seat after the soft seating rings have disintegrated. The stuffing box may be fitted with an auxiliary asbestos based or pure graphite packing, or the packing may be made entirely of an asbestos compound or pure graphite.

Numerous standards have been established which cover the requirements for testing and evaluating the performance of soft-seated ball valves when exposed to fire. The three premiere standards are:

BS 5146

API 607

API RP 6F

BS 5146 is a derivative of OCMA FS VI, which itself was taken from a test specification created by Esso Petroleum in the U.K. This test differed from all former ones in requiring the valve to be in the open position during the test and using a flammable liquid in the valve. The test owes its origin to a recognition by Esso that, in an actual plant fire, a significant number of valves may be in the open position and have to be subsequently closed. Moves are under foot between a number of standard organizations to arrive at an international fire-test specification.

Besides paying attention to the fire-testing of ball valves in systems handling flammable fluids, similar attention must be paid to the effect of fire on the entire fluid handling system, including valves other than ball valves, valve operators, pumps, filters, pressure vessels; and not least, the pipe flanges, bolting and gaskets.

Fire-tested ball valves are also referred to as fire-safe. However, this term is unacceptable to valve manufacturers from the product liability standpoint.

Multiport Configuration

Ball valves adapt to multiport configurations in a manner similar to plug valves, previously discussed on page 96.

Ball Valves for Cryogenic Service

Ball valves are used extensively in cryogenic services, but their design must be adapted for this duty. A main consideration in the design of these valves is the coefficient of thermal contraction of the seat ring material which is normally higher than that of the stainless steel of the ball and valve body. The seat rings shrink, therefore, on the ball at low temperatures and cause the operating torque to increase. In severe cases, the seat ring may be overstressed, causing it to split.

This effect of differential thermal contraction between the seats and the ball may be combatted by reducing the installed prestress between the seats and the ball by an amount which ensures a correct prestress at the cryogenic operating temperature. However, the sealing capacity of these valves may not be satisfactory at low fluid pressures if these valves also have to operate at ambient temperatures.

Other means of combatting the effect of differential thermal contraction between the seats and the ball include supporting the seats on flexible metal diaphragms; choosing a seat-ring material which has a considerably lower coefficient of contraction than virgin PTFE, such as graphite or carbon filled PTFE; or making the seat rings of stainless steel with PTFE inserts, in which the PTFE contents are kept to a minimum.

Because plastic seat-ring materials become rigid at cryogenic temperatures, the surface finish of the seatings and the sphericity of the ball must be of a high standard to ensure a high degree of seat tightness. Also, as with other types of valves for cryogenic service, the extended bonnet should be positioned no more than 45° from the upright to ensure an effective stem seal.

Variations of Body Construction

Access to the ball valve internals can be provided in various ways. This has led to the development of a number of variations in the body construction; Figure 3-63 shows the most common variations.

The one-piece body has the fewest number of body joints subject to pipeline stresses. This type of body is therefore often chosen for hazardous fluids. If the valve is to be buried, the sealed-body variety is frequently used. The one-piece body with top entry and the various split body constructions offer easy entry to the valve internals. In the case of welded-in-line valves, those with top entry may also be serviced *in situ*. A selection from these types is often a matter of personal preference.

Face-to-Face Dimensions

The original practice of USA manufacturers was to make the face-to-face dimensions of flanged ball valves to the nearest valve standard which gave minimum material content. This happened to be the gate valve standard, but the face-to-face dimensions of class 150, and of sizes DN 200 (NPS 8) through DN 300 (NPS 12) of class 300, permitted only reduced-bore construction.

In 1961, when UK manufacturers also introduced the flanged ball valve, there was an additional demand for full-bore ball valves. Where it was impossible to accommodate the full-bore ball valve in the confines of the face-to-face dimensions of gate valves, the face-to-face dimensions of regular-pattern plug valves were adopted.

Thus, there is a short and a long series of ball valves for class 150, and in sizes DN 200 (NPS 8) through DN 300 (NPS 12), also for class 300—one for

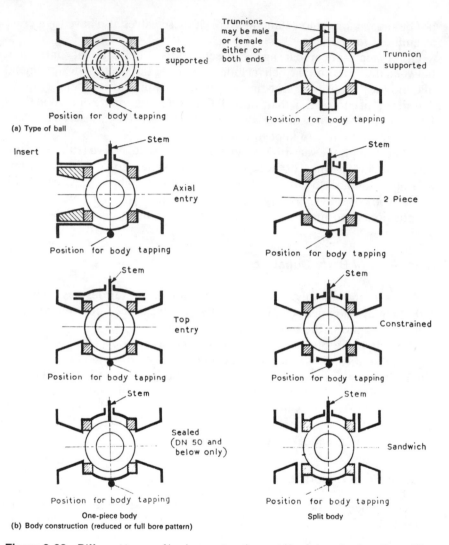

Figure 3-63. Different types of body construction, and body tapping locations. (Reprinted from BS 5159: 1974, courtesy of British Standards Institution.)

reduced-bore and one for full-bore ball valves, respectively. In the case of the higher pressure ratings, the face-to-face dimensions of gate valves accommodate both reduced-bore and full-bore ball valves throughout.

The master standard for face-to-face dimensions is ISO 5752. This standard includes all the recognized dimensions worldwide that are used in the piping industry. However, ISO 5752 does not try to define reduced-bore or full-bore except for sizes DN 200 (NPS 8) through DN 300 (NPS 12) of class 300.

Standards Pertaining to Ball Valves

A list of USA and British standards pertaining to ball valves may be found in Appendix C.

Applications

Duty:
  Stopping and starting flow.
  Moderate throttling.
  Flow diversion.
Service:
  Gases.
  Liquids.
  Non-abrasive slurries.
  Vacuum.
  Cryogenic.

BUTTERFLY VALVES

Butterfly valves are rotary valves in which a disc-shaped closure member is rotated through 90°, or about, to open or close the flow passage. The butterfly valves shown in Figures 3-64 through 3-76b represent a cross section of the many variations available.

The original butterfly valve is the simple pipeline damper which is not intended for tight shut-off. This valve is still an important member of the butterfly valve family.

The advent of elastomers has initiated the rapid development of tight shut-off butterfly valves in which the elastomer serves as the sealing element between the rim of the disc and the valve body. The original use of these valves was for water.

As more chemical-resistant elastomers became available, the use of butterfly valves spread to wide areas of the process industries. The elastomers used for these purposes must not only be corrosion resistant but also abrasion resistant, stable in size, and resiliency retentive, i.e., they must not harden. If one of these properties is missing, the elastomer may be unsuitable. Valve manufacturers can advise on the selection and limitations of elastomers for a given application.

Efforts to overcome some of the limitations of elastomers led to the development of butterfly valves with PTFE seats. Other efforts led to the development of tight shut-off butterfly valves with metal seatings. By these developments, butterfly valves became available for a wide range of pressures and temperatures, based on a variety of sealing principles.

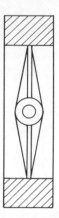 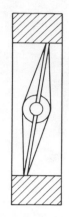 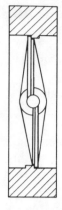

swing-thru disc
and metal seating,
not fluid tight

angle-seated disc
and metal seating,
not fluid tight

step-seated disc
and metal seating,
not fluid tight

Figure 3-64. Butterfly valves. (Courtesy of GEC-Elliot Control Valves Limited.)

Figure 3-65. Butterfly valve with resilient replaceable liner and interference-
seated disc. (Courtesy of Keystone International, Inc.)

Butterfly valves give little resistance to flow when fully open, and also
sensitive flow control when open between about 15° and 70°. Severe throttling of liquids may, of course, lead to cavitation, depending on the vapor
pressure of the liquid and the downstream pressure. Any tendency of the
liquid to cavitate may be combatted partly by sizing the butterfly valve
smaller than the pipeline so that throttling occurs about in the half-open

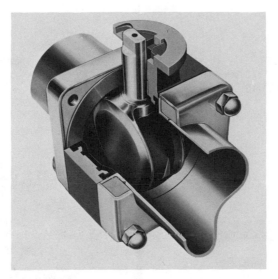

Figure 3-66. Butterfly valve with resilient liner and interference-seated double disc for double block and bleed, used for isolating food stuffs from cleaning-in-place cleaning fluids. (Courtesy of Amri SA.)

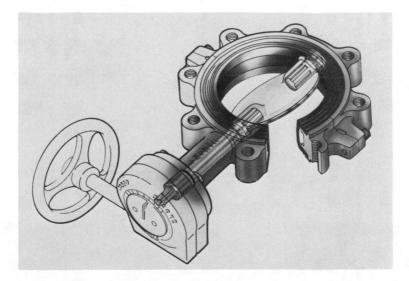

Figure 3-67. Butterfly valve with resilient replaceable liner and interference-seated disc, liner bonded to steel band, valve body split in two halves. (Courtesy of DeZurik.)

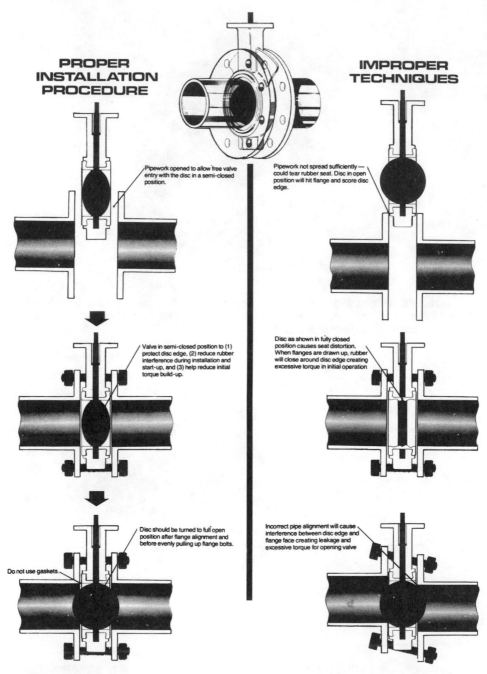

Figure 3-68. Proper and improper installation procedures for interference-seated butterfly valves with replaceable rubber liner. (Courtesy of Keystone International, Inc.)

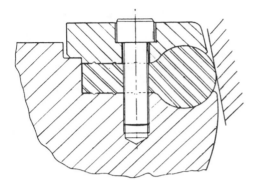

Figure 3-69. Scrap view of interference-seated butterfly valve, showing the resilient sealing element carried on the rim of the disc. (Courtesy of Boving & Co., Limited.)

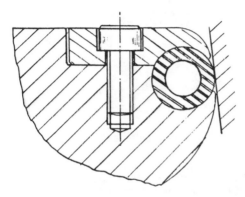

Figure 3-70. Scrap view of butterfly valve, showing an inflatable sealing element carried on the rim of the disc. (Courtesy of Boving & Co., Limited.)

position, and/or by letting the pressure drop occur in steps, using a number of valves, as discussed in Chapter 2, page 35. Also, if the butterfly valve is closed too fast in liquid service, waterhammer may become excessive. By closing the butterfly valve slowly, excessive waterhammer can be avoided, as discussed in Chapter 2, page 36.

Because the disc of butterfly valves moves into the seat with a wiping motion, most butterfly valves are capable of handling fluids with solids in suspension and, depending on the robustness of the seatings, also powders and granules. If the pipeline is horizontal, the butterfly valve should be mounted with the stem in the horizontal and rotated so that the bottom of the disc lifts away from the solids in the direction of flow.

Seating Designs

From the point of seat tightness, butterfly valves may be divided into nominal-leakage valves, low-leakage valves, and tight shut-off valves. The nominal- and low-leakage valves are used mainly for throttling or flow

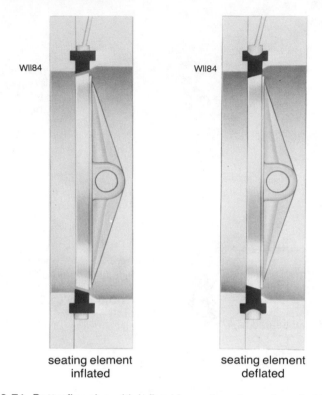

seating element
inflated

seating element
deflated

Figure 3-71. Butterfly valve with inflatable sealing element carried in a recess of the valve body. (Courtesy of GEC-Elliot Control Valves, Limited.)

control duty, while tight shut-off butterfly valves are used for tight shut-off, throttling, or flow control duty.

The butterfly valves shown in Figure 3-64 are examples of nominal- and low-leakage butterfly valves in which both the seat and disc are metallic. For applications in which a lower leakage rate is required, butterfly valves are available in which the rim of the disc is provided with a piston ring.

The intimate contact between the seatings of tight shut-off butterfly valves may be achieved by various means. Some of the ones more frequently used are:

1. By interference seating which requires the disc to be jammed into the seat.
2. By forcing the disc against the seat, requiring the disc to tilt about a double off-set hinge in a manner comparable to tilting disc check valves, as described on page 150.

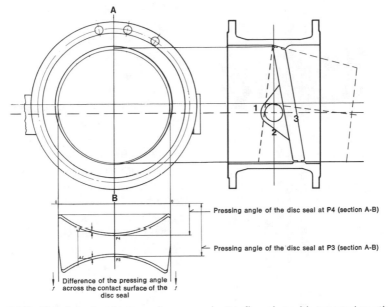

Figure 3-72. Metal-seated high performance butterfly valve with tapered seating faces and seating of the disc at a non-interlocking angle. (Courtesy of Clow Corporation.)

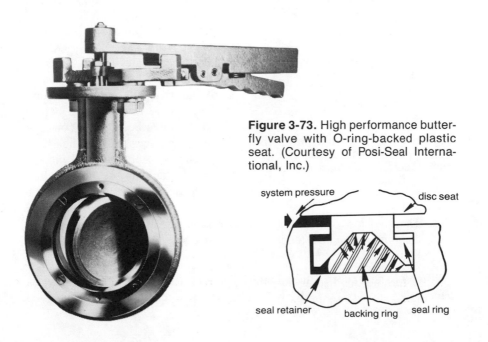

Figure 3-73. High performance butterfly valve with O-ring-backed plastic seat. (Courtesy of Posi-Seal International, Inc.)

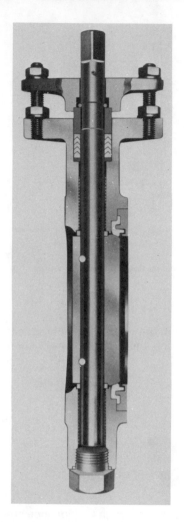

Figure 3-74. High performance butterfly valve with flexible PTFE lip seat. (Courtesy of Jamesbury International Corp.)

3. By pressure-energized sealing, using sealing elements such as O-rings, lip seals, diaphragms and inflatable hoses.

The majority of butterfly valves are of the interference seated type in which the seat is provided by a rubber liner, as in the valves shown in Figures 3-65 through 3-67. Where rubber is incompatible with the fluid to be sealed, the liner may be made of PTFE, which is backed-up by an elastomer cushion to impart resiliency to the seat.

There are a number of rubber liner constructions in common use. Typical constructions are:

1. The rubber liner may consist of a U-shaped ring that is slipped over the body without bonding, as in the valve shown in Figure 3-65. Such seats are readily replaceable. If the liner is made of a relatively rigid material such as PTFE, the valve body is split along the centerline as shown in Figure 3-67 to permit the liner to be inserted without manipulation.
2. The rubber liner may be bonded to the valve body. This construction minimizes the wearing effects of rubber bunching from disc rotation which may occur if the liner is loose. On the debit side, the rubber liner cannot be replaced.
3. The U-shaped rubber liner may be bonded to a metal band and this combination be inserted into a split valve body, as in the valve shown in Figure 3-67.

To ensure that rubber-lined butterfly valves achieve their full sealing capacity, they must be correctly installed. Two requirements must be observed:

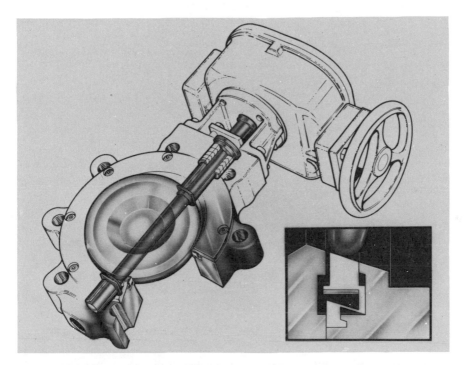

Figure 3-75. High performance butterfly valve with PTFE lip seal supported by titanium metal ring. (Courtesy of DeZurik.)

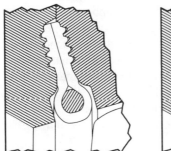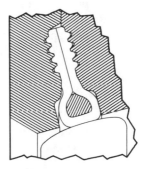

Seat not compressed as
disc approaches.

Disc in closed position;
no line pressure applied.

Disc in closed position;
line pressure applied.
Line pressure ⟶

Figure 3-76. Scrap view of high performance butterfly valve showing PTFE encapsulated elastomeric O-ring seat. (Courtesy of Rockwell International.)

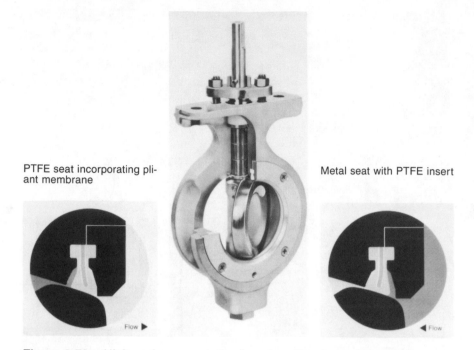

PTFE seat incorporating pliant membrane

Metal seat with PTFE insert

Flow ▶

◀ Flow

Figure 3-76a. High performance butterfly valve. (Courtesy of Xomox Corporation.)

1. The rubber liner should be fully supported by the pipe flanges. The pipe flanges should therefore be of the weld-neck type rather than the slip-on type. In the case of slip-on flanges, the rubber liner remains unsupported between the valve bore and the outside diameter of the pipe. This lack of support of the liner tends to promote distortion of the liner during valve operation, resulting in early wear and seat leakage.
2. When installing the valve, the disc must initially be put into the near-closed position so as to protect the rim of the disc from damage during handling. Prior to tightening the flange bolts, the disc must be rotated into the fully open position to permit the liner to find its undisturbed position.

Figure 3-68 illustrates the precautions that must be taken when installing rubber-lined butterfly valves.

The flange of the rubber liner also serves as a sealing element against the pipeline flanges. The installation of additional rubber gaskets between pipe flanges and valve would tend to reduce the support of the rubber liner and, consequently, reduce the sealing capacity of the valve.

The butterfly valve shown in Figure 3-69 is an interference-seated butterfly valve that carries the sealing element on the rim of the disc. The sealing element consists in this case of a heavy section O-ring with a tail clamped to the disc. By adjusting the clamping force, the seating interference can be adjusted, within limits. Because the sealing element deforms against a wide face instead of around the narrow face of a disc, the seating and unseating torques are correspondingly lower. This particular make of valve is made in larger sizes only, and for relatively high fluid pressures.

The butterfly valves shown in Figures 3-70 and 3-71 rely for a seat seal on the fluid pressure to expand a rubber sealing element against the mating seating face. The fluid pressure may be provided by the system fluid or the fluid of an external source.

The sealing element of the seating arrangement shown in Figure 3-70 consists of an inflatable hose mounted on the rim of the disc. The hose is reinforced inside by a metallic conduit, and connected through the operating shaft to the upstream system or an external fluid system. The disc is moved

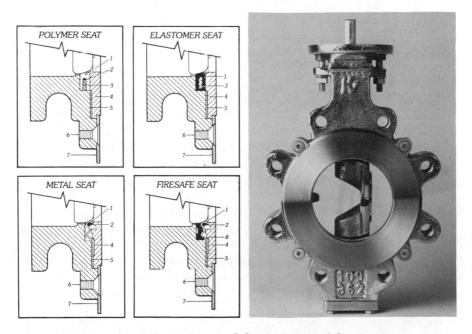

LEGEND: 1. Seat 2. Seat backing wire 3. Seat backing ring 4. Seat retaining ring
5. Seat retaining gasket 6. Flat head screw 7. Locating plate 8. RTFE member

Figure 3-76b. High performance butterfly valve designed for interference seating. Scrap views show plastic seat, elastomer seat, metal seat, and plastic/metal composition seat for flammable liquid service. (Courtesy of Keystone International, Inc.)

into the tapered seat with the hose deflated so that the seating torque is minimal. The hose is then pressurized to provide a fluid-tight seal against the seat. If the seal requires further tightening, the hose may be pumped up using a hand pump. When the valve is to be opened, the hose is first deflated so that the valve opens with a minimum of unseating torque. The valve is made to the largest sizes in use.

The sealing element of the butterfly valve shown in Figure 3-71 consists of a tubular shaped diaphragm of T-cross section which is mounted in a slot of the valve body and sealed against the flow passage. The diaphragm is pressurized on closing against the rim of the disc and depressurized on opening in a manner similar to the valve shown in Figure 3-70.

Efforts to adapt butterfly valves to wider temperature and pressure ranges have led to the development of a family of butterfly valves that may be fitted with seatings of a variety of construction materials to meet the operational requirements. Such seatings may be metal-to-polymer or metal-to-metal, and may be designed to satisfy the requirements of fire-tested valves. The majority of these valves may be used for flow in both directions.

Valves of this performance class have acquired the name *high performance butterfly valves*. The name is taken to mean that this species of butterfly valve has a greater pressure-temperature envelope than the common elastomer lined or seated butterfly valves. Figures 3-72 through 3-76b show examples of such valves. However, the illustrations do not show all the seal variations available for each particular valve.

Figure 3-72 illustrates a high performance butterfly valve in which the seat seal is achieved by forcing the disc against the seat about a double off-set hinge in a manner comparable to tilting disc check valves. The shape of the disc represents a slice from a tapered plug made at an oblique angle to the plug axis. The seating faces formed in this way are tapered on an elliptical circumference. The disc rotates around a point below the centerline of the valve and behind the seat face so that the disc drops into and lifts out of the seat with little rubbing action. By this concept, the seating load is provided by the applied closing torque from the valve operator and the hydrostatic torque from the fluid pressure acting on the unbalanced area of the disc. The valve may be provided with metal-to-metal seatings as illustrated, or the disc may be provided with a variety of soft sealing elements which seal against the metal seat provided by the body. With this choice of seating constructions, the valve may be used for differential pressures up to 3.5 MPa (500 lb/in.^{2}) and operating temperatures between $-200°C$ ($-350°F$) and $650°C$ ($1200°F$).

The concept of the high performance butterfly valve shown in Figure 3-73 may be likened to that of a ball valve which uses a wafer section of the sphere only in connection with a single seat. The illustrated seat consists of a U-shaped plastic seal ring which is mounted in a T-slot of the valve body and backed-up by an elastomeric O-ring. The O-ring imparts some initial compression stress to the seat. As the disc moves into the seat with slight interference, fluid pressure acting on the O-ring forces the seat ring into closer contact with the rim of the disc. The valve is made with a variety of seat ring constructions, including metal construction to suit a variety of operating conditions.

The concept of the high performance butterfly valve shown in Figure 3-74 is likewise borrowed from the ball valve, but with the difference from the valve shown in Figure 3-73 that the axis of the disc is offset not only from the plane of the seat, but also by a small amount from the centerline of the valve. In this way, the disc moves into the seat in a camming action, thereby moving progressively into intimate contact with the seat. Conversely, the seatings rapidly disengage during opening so that there is minimum rubbing between the seatings. The seat as illustrated consists of a PTFE lip seal with tail which is clamped between the valve body and a retainer. Once the disc has entered the seat with slight interference, fluid pressure forces the lip of the seat into more intimate contact with the rim of the disc, irrespective of whether the seat is located on the upstream or downstream side of the disc. This ability of the seatings to seal in both directions is aided by a small but controlled amount of axial movement of the disc. The valve is also available with metal and fire-tested seat constructions.

The principle of the double offset location of the disc axis in combination with pressure-energized sealing in both directions applies also to the high performance butterfly valves shown in Figures 3-75 through 3-76a.

The high performance butterfly valve shown in Figure 3-75 contains, like the valve shown in Figure 3-74, a PTFE lip seal for the seat, but of a modified shape in which the lip is supported by a titanium ring. The seat is also available in metal and fire tested construction.

The high performance butterfly valve shown in Figure 3-76 utilizes an elastomeric O-ring for the seat which is encapsuled in TFE and anchored in the valve body. The O-ring imparts elasticity and resiliency to the seat while the TFE envelope protects the O-ring from the effects of the system fluid. The seat is also available in fire-tested construction.

The high performance butterfly valve shown in Figure 3-76a incorporates a seat that is combined with a flexible membrane designed to return the seat to its original position each time the valve has been operated. One seat version consists of a T-shaped PTFE member which embeds a pliant membrane. The second fire-tested version consists of a U-shaped metal member with a PTFE insert that is carried on a metal membrane. The PTFE insert provides the seat seal under normal operating conditions. In the case of the PTFE insert being destroyed in a fire, the metal seat takes over the sealing function.

The high performance butterfly valve shown in Figure 3-76b differs from the previously described high performance butterfly valve in that it relies on interference seating for seat tightness, using a double offset disc. The seat is available either in plastic, elastomer, metal, or plastic-metal fire-tested construction to suit the duty for which the valve is to be used. The performance of the plastic and metal seats relies on a back-up wire winding which is designed to provide absolute seat rigidity when the valve is closed but to avoid imposing a load on the seat when the valve is open, thereby permitting the seat to flex radially. The rigidity in the closed position is required to achieve the desired seat interference for a tight shut-off against high and low pressures, including vacuum.

Butterfly Valves for Fire Exposure

Butterfly valves, which may be exposed to plant fire when handling flammable fluids, must remain essentially fluid-tight internally and externally and be operable during and after a fire. These conditions may be met by fire-tested versions of high performance butterfly valves.

The requirements for testing and evaluating the performance of butterfly valves when exposed to fire are similar to those for ball valves, described on page 105.

Body Configurations

The preferred body configuration of butterfly valves is the wafer, which is clamped between two pipeline flanges. An important advantage of this construction is that the bolts pulling the mating flanges together carry all the tensile stresses induced by the line strains and put the wafer in compression. This compressive stress is eased by the tensile stresses imposed by the internal fluid pressure. Flanged bodies, on the other hand, have to carry all the tensile stresses imposed by the line strains, and the tensile stresses from the line pressure are cumulative. This fact, together with the ability of most metals to handle compressive loads of up to twice their limit for tensile loads, strongly recommends the use of the wafer body.

However, if the downstream side of the butterfly valve serves also as a point of disconnection while the upstream side is still under pressure, the cross-bolted wafer body is unsuitable unless provided with a false flange. A flanged body or a lugged wafer body in which the lugs are threaded at each end to receive screws from the adjacent flanges is commonly used.

Torque Characteristic of Butterfly Valves

The torque required to operate butterfly valves consists of three main components:

T_h = hydrodynamic torque which is created by the flowing fluid acting on the disc

T_b = torque to overcome bearing friction

T_s = torque required to seat or unseat the disc

The hydrodynamic torque varies with the valve opening position and the pressure drop across the valve. In the case of symmetrical discs, this torque is identical for either direction of flow, and its direction of action is against the opening motion throughout. If the disc is offset, as in the disc shown in Figure 3-77, the hydrodynamic torque differs for each direction of flow, and the lowest torque develops when the flow is toward the disc. With flow toward the shaft, the torque acts against the opening motion throughout. However, with flow toward the disc, the torque acts only initially against the opening motion and then, with further valve opening, changes its directions.

The bearing, seating, and unseating torques, on the other hand, always act against the operating motion. The magnitude of the bearing torque corresponds to the resultant hydrodynamic force on the disc, while the magnitude of the seating and unseating torques is independent of flow.

In the case of interference-seated butterfly valves of sizes up to DN 400 (NPS 16), the seating and unseating torques normally dominate, provided the flow velocity is not too high. It is sufficient in this case to size the operating gear for these torques. However, the magnitude of the seating and unseating torques is influenced by the type of fluid handled and the operating frequency. For example, if the fluid has good lubricity and the valve is operated frequently, the seating and unseating torques are lower than for fluids which have little lubricity or consist of solids, or when the valve is operated

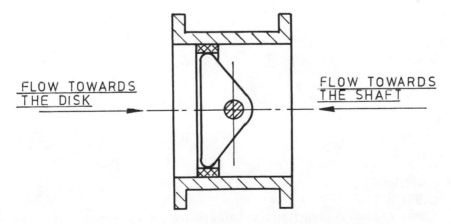

FLOW TOWARDS THE DISK

FLOW TOWARDS THE SHAFT

Figure 3-77. Offset disc configuration in butterfly valves.

infrequently. Manufacturers supply tables which give the seating and unseating torques for various conditions of operation.

In the case of interference-seated butterfly valves above DN 400 (NPS 16) and conditions of high flow velocities, the hydrodynamic and bearing torques can greatly exceed the seating and unseating torques. The operator for such valves may therefore have to be selected in consultation with the manufacturer.

Figure 3-78 shows typical opening torque characteristics of butterfly valves with symmetrical and offset discs representing the summary of torques at a constant pressure loss across the valve. Under most operating conditions, however, the pressure drop across the valve decreases as the valve is being opened, as shown for example in Figure 3-79 for a given pumping installation. The decreasing pressure drop, in turn, decreases the corresponding operating torque as the valve opens. The maximum operating torque shifts thereby from the region of the fully open valve to the region of the partially open valve, as shown in Figure 3-80. This torque is lowest if the disc is of the offset type and the direction of flow is toward the disc. However, this advantage is obtained at the penalty of a somewhat higher pressure drop. If the valve is large and the flow velocity high, this torque characteristic can be exploited to lower the cost of the operator.

Standards Pertaining to Butterfly Valves

A list of USA and British standards pertaining to butterfly valves may be found in Appendix C.

Applications

Duty:
  Stopping and starting flow.
  Controlling flow.

Service:
  Gases.
  Liquids.
  Slurries.
  Powder.
  Granules.
  Sanitary handling of pharmaceuticals and food stuffs.
  Vacuum.

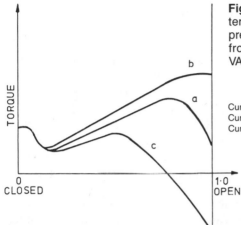

Figure 3-78. Opening torque characteristics of butterfly valves at constant pressure drop across valve. (Reprinted from Schiff and Hafen,[6] courtesy of VAG-Armaturen GmbH.)

Curve a: symmetrical disk, flow from either direction
Curve b: off-set disk, flow towards shaft
Curve c: off-set disk, flow towards disk

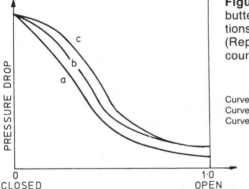

Figure 3-79. Pressure drop across butterfly valves for all opening positions in an actual pumping installation. (Reprinted from Schiff and Hafen,[6] courtesy of VAG-Armaturen GmbH.)

Curve a: symmetrical disk, flow from either direction
Curve b: off-set disk, flow towards shaft
Curve c: off-set disk, flow towards disk

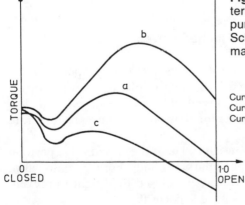

Figure 3-80. Opening torque characteristics of butterfly valves in an actual pumping installation. (Reprinted from Schiff and Hafen,[6] courtesy of VAG-Armaturen GmbH.)

Curve a: symmetrical disk, flow from either direction
Curve b: off-set disk, flow towards shaft
Curve c: off-set disk, flow towards disk

PINCH VALVES

Pinch valves are flex-body valves consisting of a flexible tube which is pinched either mechanically, as in the valves shown in Figures 3-81 through 3-83; or by the application of a fluid pressure to the outside of the valve body, as in the valves shown in Figures 3-84 and 3-85.

One of the principal advantages of this design concept is that the flow passage is straight without crevices and moving parts. The soft valve body has also the ability to seal around trapped solids. Pinch valves are therefore suitable for handling slurries and solids which would clog in obstructed flow passages, and for the sanitary handling of food stuffs and pharmaceuticals. Depending on the construction material used for the valve body, pinch valves will also handle severely abrasive fluids and most corrosive fluids.

Open and Enclosed Pinch Valves

Mechanically pinched valves may be of the open type, as in the valve shown in Figure 3-81; or of the enclosed type, as in the valves shown in Figures 3-82 and 3-83.

The open construction is used if it is desirable to inspect the valve body visually and physically while the valve is in service. When bulges and soft spots appear in the valve body, it is a sign that the valve body needs to be replaced.

If the escape of the fluid from an accidentally ruptured valve body cannot be tolerated, the valve body must be encased. The casing is normally split along the axis of the flow passage for convenient access, but casings of unit construction are also being made.

Flow Control with Mechanically Pinched Valves

Pinch valves give little flow control between the fully open and the 50% pinched position because of the negligible pressure drop at these valve positions. Any further closing of the valve gives good flow control. Mechanically pinched valves for flow control duty are therefore often 50% prepinched.

If the fluid is severely erosive, flow control near the closed valve position must be avoided to prevent grooving of the valve body. For this reason also, pinch valves for erosive duty must always be closed fluid tight to prevent grooving of the valve body as a result of leakage flow.

Figure 3-81. Pinch valve, open construction. (Courtesy of Flexible Valve Corporation.)

Figure 3-82. Pinch valve, enclosed construction, valve body 50% prepinched for sensitive flow control. Drain plug may be removed for vacuum connection when valve is used on vacuum service. (Courtesy of RKL Controls, Inc.)

Figure 3-83. Pinch valve, enclosed construction, PFTE valve body fitted with tear drops. (Courtesy of Resistoflex Corporation.)

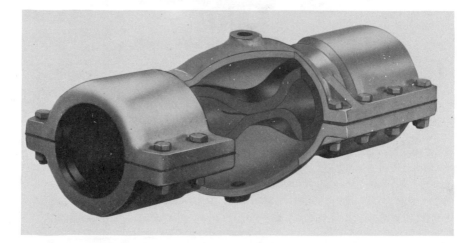

Figure 3-84. Pinch valve, fluid-pressure operated. (Courtesy of RKL Controls, Inc.)

THROTTLING ACTION

SIDE VIEW   END VIEW

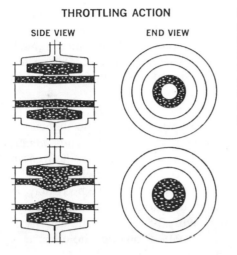

Figure 3-85. Pinch valve, fluid-pressure operated, with iris-like closing action. The constricting action of the rubber muscle upon the sleeve is literally a 360° squeeze. Pressure is evenly applied to the circumference of the sleeve—hence the always round, always centered hole. (Courtesy of The Clarkson Company.)

Flow Control with Fluid-Pressure Operated Pinch Valves

Fluid-pressure operated pinch valves are not normally suitable for manual flow control because any change in the downstream pressure will automatically reset the valve position. The flow control characteristic of fluid-pressure operated pinch valves, such as the one shown in Figure 3-84, is otherwise similar to that of mechanically pinched valves.

An exception to this flow control characteristic is the valve shown in Figure 3-85, in which the closing action of the valve body is iris-like. This closing action gives equal-percentage flow control throughout.

The iris-like closure allows the valve to pass larger particles through in any valve position than is possible with any other valve. This ability considerably reduces the tendency of some slurries to bridge in the partly closed valve. The circular throttling orifice tends to eliminate also the grooving of the valve body which develops in other types of pinch valves when severely throttling abrasive fluids. On the debit side, the valve cannot be fully closed.

Valve Body

The valve body is made of an elastomer, PTFE or a similar material, or a combination of these materials. Fabric reinforcement may be used to increase the mechanical strength of the valve body.

Natural rubber is the best construction material for the valve body from the point of resistance to cracking from flexing. Natural rubber also shows excellent resistance to abrasive wear and to corrosion from many corrosive fluids.

If the fluid is chemically active, the surface of highly stressed rubber is more readily attacked than that of unstressed rubber. The corrosives may also form a thin, highly resistant film of corrosion products which is much less flexible than the parent rubber. This film will crack on severe flexing and expose the rubber to further attack. Repeated flexing will finally lead to cracking right through the rubber.

Some of the synthetic rubbers are considerably less subject to this form of attack than natural rubber. These rubbers extend, therefore, the application of pinch valves for corrosive fluids. In developing such compounds, manufacturers aim not only at corrosion resistance but also at high tensile strength combined with softness for abrasion resistance and ease of closure.

To minimize the severity of flexing, the inside of the valve body may be provided with groove-like recesses on opposite sides, along which the valve body folds on closing. The valve shown in Figure 3-83 reduces the severity of flexing by providing tear drops inside the valve body on opposite sides. This design permits the valve body to be made entirely of PTFE.

Limitations

The flexible body puts some limitations on the use of pinch valves. These limitations may be overcome in some cases by special body designs or by the method of valve operation.

For example, fluid-pressure operated pinch valves tend to collapse on suction duty. If mechanically pinched valves are used for this duty, the body must be positively attached to the operating mechanism.

Pinch valves on the downstream side of a pump should always be opened prior to starting up the pump. If the valve is closed, the air between the pump and the valve will compress and, upon cracking the valve, escape rapidly. The liquid column which follows will then hit the valve body with a heavy blow — perhaps severe enough to burst the valve body.

Pinch valves may also fail if the flow pulsates. The valve body will pant and finally fail due to fatigue.

There is a limitation when using pinch valves as the main shut-off valve in liquid-handling systems in which the liquid can become locked and has no room to move. Because the valve must be able to displace a certain amount of liquid when closing, the valve cannot be operated in this situation. Any effort to do so may cause the valve body to burst.

Standards Pertaining to Pinch Valves

Refer to Appendix C for standards pertaining to pinch valves.

Applications

Duty:
  Stopping and starting flow.
  Controlling flow.

Service:
  Liquids.
  Abrasive slurries.
  Powders.
  Granules.
  Sanitary handling of pharmaceuticals and food stuffs.

DIAPHRAGM VALVES

Diaphragm valves are flex-body valves in which the valve body consists of a rigid and flexible section. The flexible body section is provided by a diaphragm which, in connection with a compressor, represents the closure member. The seat is provided by the rigid body section and may consist of a weir across the flow passage, as in the valve shown in Figure 3-86, or be provided by the wall of a straight-through flow passage, as in Figure 3-87.

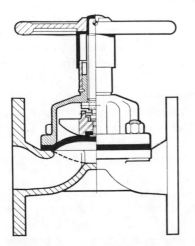
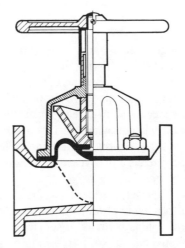

Figure 3-86. Diaphragm valve, weir type. (Courtesy of Saunders Valve Company Limited.)

Figure 3-87. Diaphragm valve, straight-through type. (Courtesy of Saunders Valve Company Limited.)

Diaphragm valves share a similar advantage with pinch valves: namely, a flow passage that is not obstructed by moving parts and is free of crevices. They may, therefore, be put to uses similar to pinch valves, including the sanitary handling of foodstuffs and pharmaceuticals.

Weir-Type Diaphragm Valves

The weir in the flow passage is designed to reduce flexing of the diaphragm to a minimum, while still providing a smooth and streamlined flow passage. The flexing stress in the diaphragm is therefore minimal, resulting in a correspondingly long diaphragm life. The short stroke of these valves also permits the use of plastics such as PTFE for the diaphragm which would be too inflexible for longer strokes. The back of such diaphragms is lined with an elastomer which promotes a uniform seating stress upon valve closing.

Weir-type diaphragm valves may also be used in general and high vacuum service. However, some valve makes require a specially reinforced diaphragm in high vacuum service.

Because the diaphragm area is large compared with the flow passage, the fluid pressure imposes a correspondingly high force on the raised diaphragm. The resulting closure torque and time limit the size to which diaphragm valves can be made. Typically, weir-type diaphragm valves of the type shown in Figure 3-86 are made in sizes up to DN 350 (NPS 14). Larger weir-type diaphragm valves up to DN 500 (NPS 20) are provided with a double bonnet assembly, as shown in Figure 3-87a.

Weir-type diaphragm valves are also available with a body of Tee configuration, as shown in Figure 3-87b, in which a branch connects to the main flow passage without impeding pipeline flow. The main function of these valves is for sampling duty where the taking of a true sample from the flowing fluid must be assured.

The stem of the valve shown in Figure 3-86 is of the rising type. To protect the external stem thread from dust and immediate outside corrosive influences, the handwheel carries a shroud that covers the stem thread while sliding over a yellow lift-indicator sleeve. The yellow lift-indicator sleeve, in turn, carries a prepacked lubrication chamber to lubricate the stem thread for long life.

If required, the stem may be provided with an O-ring seal against the bonnet to prevent fluid from escaping into the surroundings of the valve should the diaphragm break in service.

Conventional weir-type diaphragm valves may also be used in horizontal lines which must be self draining. Self draining is achieved by mounting the valve with the stem approximately 15° to 25° up from the horizontal, provided the horizontal line itself has some fall.

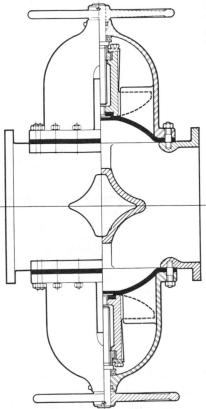

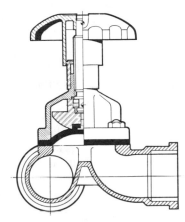

Figure 3-87a. Diaphragm valve, weir type, with double-bonnet assembly in connection with large valve sizes. (Courtesy of Saunders Valve Company Limited.)

Figure 3-87b. Diaphragm valve, weir type, with T-branch. (Courtesy of Saunders Valve Company Limited.)

Straight-Through Diaphragm Valves

Diaphragm valves with a straight-through flow passage require a more flexible diaphragm than weir-type diaphragm valves. For this reason, the construction material for diaphragms of straight-through diaphragm valves is restricted to elastomers.

Because of the high flexibility and large area of these diaphragms, high vacuum will tend to balloon the diaphragm into the flow passage. The degree of ballooning varies thereby with make, causing either a small and acceptable reduction in flow area only or a collapse of the diaphragm. In the latter case, the bonnet must be evacuated to balance the pressures on the diaphragm. When using these valves for high vacuum, the manufacturer should be consulted.

Straight-through diaphragm valves are also available with full bore and reduced bore flow passage. In the case of reduced bore valves, the bonnet assembly of the next smaller valve is used. For example, a DN 50 (NPS 2) reduced bore valve is fitted with a DN 40 (NPS 1 1/2) bonnet. The construction of the bonnet is otherwise similar to that of weir-type diaphragm valves.

The size range of straight-through diaphragm valves typically covers valves up to DN 350 (NPS 14).

Construction Materials

The valve body of diaphragm valves is available in a great variety of construction materials, including plastics, to meet service requirements. The simple body shape also lends itself readily to lining with a great variety of corrosion-resistant materials, leading often to low cost solutions for an otherwise expensive valve.

Because the diaphragm separates the bonnet from the flowing fluid, the bonnet is normally made of cast iron and epoxy coated inside and outside. If requested, the bonnet is available also in a variety of other materials.

Diaphragms are available in a great variety of elastomers and plastics. Valve makers' catalogs advise the user on the selection of the diaphragm material for a given application.

Valve Pressure/Temperature Relationships

Figures 3-87c and 3-87d show typical pressure/temperature relationships of weir-type and straight-through diaphragm valves. However, not all construction materials permit the full range of these relationships. The valve user must therefore consult the manufacturer's catalog for the permissible operating pressure at a given operating temperature.

Valve Flow Characteristics

Figure 3-87e shows the typical inherent and installed flow characteristics of weir-type diaphragm valves, though the shapes of the curves may vary to some extent between valve sizes. In the case of automatic control, these characteristics may be modified using a variable cam positioner.

Operational Limitations

Like pinch valves, diaphragm valves displace a certain amount of fluid when operated; i.e., diaphragm valves are positive-displacement type valves. For this reason, diaphragm valves must not be installed as shut-off valves in lines containing an incompressible fluid where the fluid has no room to move.

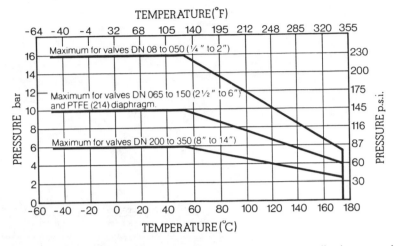

Figure 3-87c. Pressure/temperature relationship of weir-type diaphragm valves. (Courtesy of Saunders Valve Company Limited.)

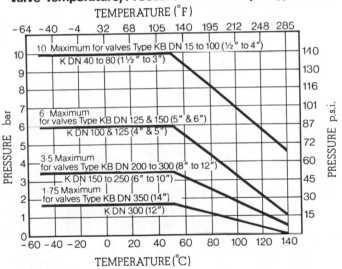

Figure 3-87d. Pressure/temperature relationship of straight-through type diaphragm valves. (Courtesy of Saunders Valve Company Limited.)

Standards Pertaining to Diaphragm Valves

A list of USA, British and ISO standards pertaining to diaphragm valves may be found in Appendix C.

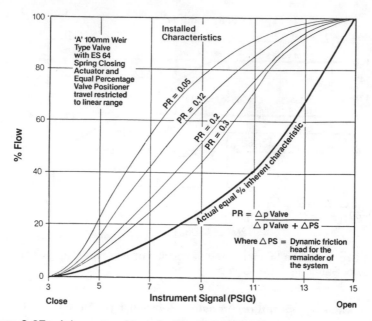

Figure 3-87e. Inherent and installed flow characteristics of weir-type diaphragm valves. (Courtesy of Saunders Valve Company Limited.)

Applications

Duty for weir-type and straight-through diaphragm valves:

Stopping and starting flow
Controlling flow

Service for weir-type diaphragm valves:

Gases, may carry solids
Liquids, may carry solids
Viscous fluids
Leak-proof handling of hazardous fluids
Sanitary handling of pharmaceuticals and foodstuffs
Vacuum

Service for straight-through diaphragm valves:

Gases, may carry solids
Liquids, may carry solids

Viscous fluids
Sludges
Slurries, may carry abrasives
Dry media
Vacuum (consult manufacturer)

STAINLESS STEEL VALVES[40,41]

Corrosion-Resistant Alloys

The least corrosion-resistant alloy is normally thought of as steel AISI type 304 (18 Cr, 10 Ni). Stainless steel AISI type 316 (18 Cr, 12 Ni, 2.5 Mo) has a wider range of corrosion resistance than type 304, and valve makers often endeavor to standardize on type 316 as the least corrosion-resistant alloy. If the valve is to be welded into the pipeline, the low carbon grades (less than 0.3% carbon) are better than stabilized grades. Flanged valves require welding only when a casting defect has to be repaired. Because the repair is done prior to the 1100°C (2000°F) water quench solution anneal, standard carbon grades are quite satisfactory for flanged valves.

References 19 through 24 list some of the publications which discuss the selection of stainless steel for corrosive fluids.

Crevice Corrosion

Practically all corrosion-resistant alloys are susceptible to crevice corrosion. Good valve designs therefore avoid threading any component which comes in contact with the corrosive fluid. For this reason, valve seats are normally made an integral part of the valve body. An exception are body designs in which the seat is clamped between two body halves, as in the valve shown in Figure 3-11. However, the gaskets between the seat and the body halves must be cleanly cut to avoid crevices.

If the valve is to be screwed into the pipeline, seal welding will improve the performance of the screwed connection. Alternatively, thread sealants which harden after application are helpful in combatting crevice corrosion in threaded joints. Flanged facings which incorporate crevices, such as tongue and groove, should be avoided.

The points of porosity in the valve body which are exposed to the corrosive fluid can likewise produce crevice corrosion. The body may thereby corrode through at the point of porosity and produce gross leakage, while the remainder of the body stays in good condition.

Galling of Valve Parts

Published information usually shows that stainless steel in sliding contact — particularly austenitic grades of like composition — are susceptible to galling. This galling tendency diminishes considerably if the fluid has good lubricity and the seating surfaces can retain the lubricants and protective contaminants. Polished surfaces have only a limited ability to retain lubricants and contaminants and therefore display an increased tendency to gall when in sliding contact. For this reason, the seating faces of stainless steel valves are usually very finely machined rather than polished.

If a high galling tendency is expected, as when handling dry gases, one of the seating faces may be faced with stellite, which is known to provide good resistance to a wide range of corrosives.

Seizing of the valve stem in the yoke bush is commonly avoided by choosing dissimilar materials for the yoke bush and the stem. A material frequently used for the bush in stainless steel valves is Ni-Resist ductile iron D2, which provides complete freedom from galling due to the graphite in its structure. If the bush is made of stainless steel, the free-machining grade type 303 provides remarkable freedom from galling in conjunction with stainless steel grades type 304 and 316 for the stem.

Light-weight Valve Constructions

Efforts in the USA to reduce the cost of stainless steel valves led to the development of standards for 150 lb light-weight stainless steel valves. The pressure ratings specified in these standards apply only to valves made from austenitic materials.

The flanges to these standards are thinner than the corresponding full-rating carbon steel flanges and have plain flat faces. Many users object to the light-weight flanges and request full-rating flanges with a raised face.

The light-weight bodies are, of course, more flexible than the bodies of full-rating carbon steel valves. This is particularly important for gate valves, in which body movements can unseat the disc. Experience has shown that plain solid wedges may be used only for sizes up to DN 100 (NPS 4). Larger valves are satisfactory only if the wedge is of the self-aligning type.

Standards Pertaining to Stainless Steel Valves

A list of standards pertaining to stainless steel valves may be found in Appendix C.

4

Check Valves

Function of Check Valves

Check valves are automatic valves which open with forward flow and close against reverse flow.

This mode of flow regulation is required to prevent return flow, to maintain prime after the pump has stopped, to enable reciprocating pumps and compressors to function, and to prevent rotary pumps and compressors from driving standby units in reverse. Check valves may also be required in lines feeding a secondary system in which the pressure can rise above that of the primary system.

Grouping of Check Valves

Check valves may be grouped according to the way the closure member moves onto the seat. Four groups of check valves are then distinguished:

1. Lift check valves. The closure member travels in the direction normal to the plane of the seat, as in the valves shown in Figures 4-1 through 4-6a.
2. Swing check valves. The closure member swings about a hinge which is mounted outside the seat, as in the valves shown in Figures 4-7 and 4-8.
3. Tilting-disc check valves. The closure member tilts about a hinge which is mounted near, but above, the center of the seat, as in the valve shown in Figure 4-9.
4. Diaphragm check valves. The closure member consists of a diaphragm which deflects from or against the seat, as in the valves shown in Figures 4-10 through 4-11b.

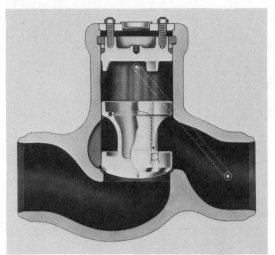

Figure 4-1. Lift check valve with piston-type disc, standard pattern. (Courtesy of Flow Control Division, Rockwell International.)

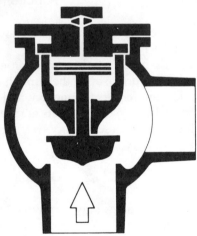

Figure 4-2. Lift check valve, angle pattern, with built-in dashpot which comes into play during the final closing movement. (Courtesy of Sempell Armaturen.)

Operation of Check Valves

Check valves shall operate in a manner which avoids:

1. The formation of an excessively high surge pressure as a result of the valve closing.
2. Rapid fluctuating movements of the valve closure member.

To avoid the formation of an excessively high surge pressure as a result of the valve closing, the valve must close fast enough to prevent the development of a significant reverse flow velocity which on sudden shut-off is the source of the surge pressure.

However, the speed with which forward flow retards can vary greatly between fluid systems. If, for example, the fluid system incorporates a number of pumps in parallel and one fails suddenly, the check valve at the outlet of the pump that failed must close almost instantaneously. On the other hand, if the fluid system contains only one pump which suddenly fails, and if the delivery line is long and the back pressure at the outlet of the pipe and the pumping elevation are low, a check valve with a slow closing characteristic is satisfactory.

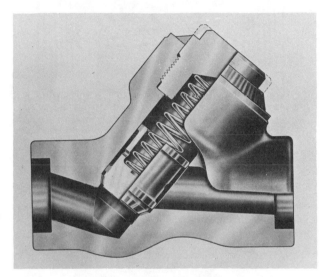

Figure 4-3. Lift check valve with piston-type disc, oblique pattern. (Courtesy of Flow Control Division, Rockwell International.)

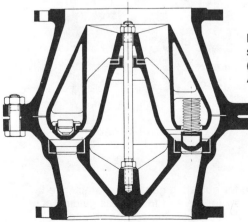

Figure 4-4. Lift check valve with spring-loaded ring-shaped disc. (Courtesy of Mannesmann-Meer AG.)

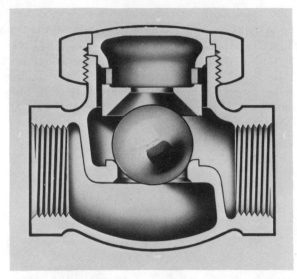

Figure 4-5. Lift check valve ball-type disc, standard pattern. (Courtesy of Crane Co.)

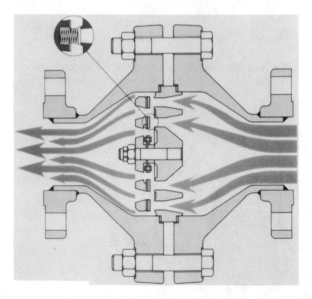

Figure 4-6. Lift check valve for pulsating gas flow, characterized by minimum valve lift, and low inertia and frictionless guiding of closure member. (Courtesy of Hoerbiger Corporation of America.)

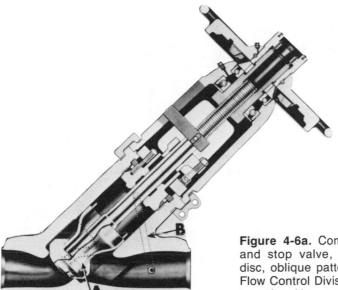

Figure 4-6a. Combined lift check and stop valve, with piston-type disc, oblique pattern. (Courtesy of Flow Control Division, Rockwell International.)

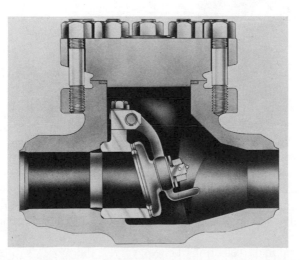

Figure 4-7. Swing check valve. (Courtesy of Velan Engineering, Limited.)

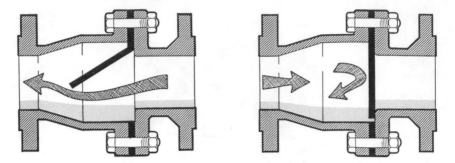

Figure 4-7a. Swing check valve with steel reinforced rubber disc, the disc being an integral part of the body gasket. (Courtesy of Saunders Valve Company Limited.)

Figure 4-8. Double-disc swing check valve. (Courtesy of Mission Manufacturing Co.)

Rapid fluctuating movements of the closure member must be avoided to prevent excessive wear of the moving valve parts which could result in early failure of the valve. Such movements can be avoided by sizing the valve for a flow velocity which forces the closure member firmly against a stop.

If flow pulsates, check valves should be mounted as far away as practical from the source of flow pulsations. Rapid fluctuations of the closure member may also be caused by violent flow disturbances. Where this situation exists, the valve should be located at a point where flow disturbances are at a minimum.

The first step in the selection of check valves, therefore, is to recognize the conditions under which the valve operates.

Assessment of Check Valves for Fast Closing[28]

In most practical applications, check valves can be assessed for fast closing speed only qualitatively. The following criteria may serve as a guide:

1. Travel of the closure member from the fully open to the closed position should be as short as possible. Thus, from the point of speed of closing,

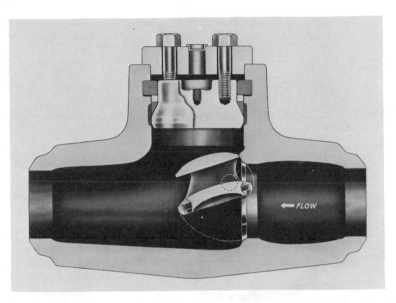

Figure 4-9. Tilting-disc check valve. (Courtesy of Flow Control Division, Rockwell International.)

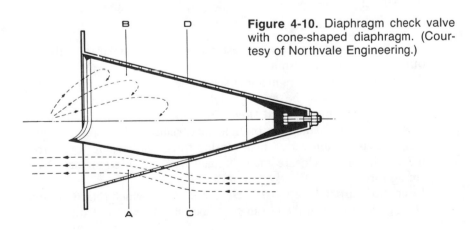

Figure 4-10. Diaphragm check valve with cone-shaped diaphragm. (Courtesy of Northvale Engineering.)

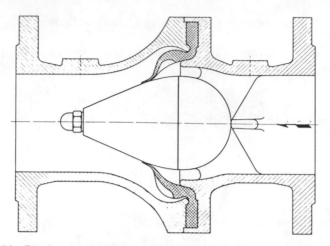

Figure 4-11. Diaphragm check valve with ring-shaped pleated diaphragm. (Courtesy of VAG-Armaturen GmbH.)

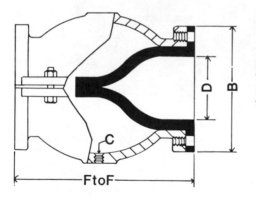

Figure 4-11a. Diaphragm check valve, incorporating flattened rubber sleeve closure member. (Courtesy of Red Valve Company, Inc.)

a smaller valve is potentially faster closing than a larger valve of otherwise the same design.

2. The inertia of the closure member should be as low as possible, but the closing force should be appropriately high to ensure maximum response to declining forward flow. From the point of low inertia, the closure member should be of light construction. To combine light-weight construction with a high closing force, the closing force from the weight of the closure member may have to be augmented by a spring force.

3. Restrictions around the moving closure member which retard the free closing movement of the closure member should be avoided.

Application of Mathematics to the Operation of Check Valves

The application of mathematics to the operation of check valves is of relatively recent origin. Pool, Porwit, and Carlton[29] describe a calculation method for check valves with a hinged disc, which involves setting-up the equation of motion for the disc and applying to that the deceleration characteristic of the flowing fluid within the system. Before the equation of motion for the disc can be written, certain physical constants of the valve must be

Figure 4-11b. Diaphragm check valve incorporating flattened rubber sleeve closure member, used as tide gate. (Courtesy of Red Valve Company, Inc.)

known. The calculation determines the reverse flow velocity at the instant of sudden shut-off. The surge pressure due to the sudden shut-off of the reverse flow can then be calculated as described in Chapter 2, page 36.

As far as the valve user is concerned, it is important for him to know that valve manufacturers can use mathematics in designing check valves for given critical applications and predicting surge pressure.

Design of Check Valves

Lift Check Valves

The check valves shown in Figures 4-1 through 4-6a represent a cross section of the family of lift check valves.

Lift check valves have an advantage over most other types of check valves in that they need only a relatively short lift to obtain full valve opening. This lift is a minimum in lift check valves in which the flow passage at the seat is ring shaped, as in the valves shown in Figures 4-4 and 4-6. Lift check valves are, therefore, potentially fast closing.

In the majority of lift check valves, the closure member moves in a guide to ensure that the seatings mate on closing. However, such guiding also has a disadvantage in that dirt entering the guide can hang-up the closure member, and viscous liquids will cause lazy valve operation or even cause the closure member to hang-up. These types of lift check valves are therefore suitable for low viscosity fluids only, which are essentially free of solids. Some designs overcome this disadvantage, as in the valve shown in Figure 4-5, in which the closure member is ball-shaped and allowed to travel without being closely guided. When the valve closes, the ball-shaped closure member rolls into the seat to achieve the required alignment of the seatings.

The check valve shown in Figure 4-2 is specifically designed for applications in which a low surge pressure is critical. This is achieved in two ways, firstly by providing the closure member with a conical extension that progressively throttles the flow as the valve closes, and secondly by combining the closure member with a dashpot that comes into play in the last closing moments. A spring to assist closing of the valve has been purposely omitted, as breakage of the spring was considered a hazard for the service for which the valve is intended.

The check valve shown in Figure 4-6 is designed for gas service only. Depending on flow conditions, the valve may serve either as a constant-flow check valve, in which case the valve remains fully open in service irrespective of minor flow fluctuations; or as a pulsating-flow check valve, in which case the valve opens and closes with each pulse of the flowing gas.

Constant-flow check valves are used after centrifugal, lobe type, and screw compressors, or after reciprocating compressors if the flow pulsations are low enough not to cause plate-flutter. Pulsating-flow check valves are used after reciprocating compressors if the flow pulsations cause the valve to open and close with each pulsation. The valves are designed on the same principles as compressor valves and, therefore, are capable of withstanding the repeated impacts between the seatings. The manufacturer advises whether a constant-flow or pulsating-flow check valve may be used for a given application.

The valves owe their operational characteristics to their design principle, based on minimum valve lift in conjunction with multiple ring-shaped seat orifices, low inertia of the plate-like closure member, frictionless guiding of the closure member, and the selection of a spring which is appropriate for the operating conditions.

The valve shown in Figure 4-6a is a combined lift check and stop valve. The valve resembles an oblique pattern globe valve in which the closure member is disconnected from the stem. When the stem is raised, the valve acts as a lift check valve. When the valve is lowered and firmly pressed against the closure member, the valve acts as a stop valve.

Lift check valves must be mounted in a position in which the weight of the closure member acts in the closing direction. Exceptions are some spring-loaded low-lift check valves in which the spring force is the predominant closing force. For this reason, the valves shown in Figures 4-1 and 4-5 may be mounted in the horizontal flow position only, while the valve shown in Figure 4-2 may be mounted in the vertical upflow position only. The valves shown in Figures 4-3, 4-4, and 4-6a may be mounted in the horizontal and vertical upflow positions, while the valve shown in Figure 4-6 may be mounted in any flow position, including vertical downflow.

Swing Check Valves

Conventional swing check valves are provided with a disc-like closure member which swings about a hinge outside the seat, as in the valves shown in Figures 4-7 and 4-7a. Travel of the disc from the fully open to the closed position is greater than in most lift check valves. On the other hand, dirt and viscous fluids cannot easily hinder the rotation of the disc around the hinge. In the valve shown in Figure 4-7a, the closure member is an integral part of the rubber gasket between the valve body halves. It is steel reinforced, and opens and closes by bending a rubber strip connecting the closure member and the gasket.

As the size of swing check valves increases, weight and travel of the disc eventually become excessive for satisfactory valve operation. For this reason, swing check valves larger than about DN600 (NPS 24) are frequently designed as multi-disc swing check valves, and have a number of conventional swing discs mounted on a multi-seat diaphragm across the flow passage in the valve.

Swing check valves should be mounted in the horizontal position, but may also be mounted in the vertical position, provided the disc is prevented from reaching the stalling position. In the latter case, however, the closing moment of the disc due to its weight is very small in the fully open position, so the valve will tend to close late. To overcome slow response to retarding flow, the disc may be provided with a lever-mounted weight or spring loaded.

The check valve shown in Figure 4-8 is a double-disc swing check valve which is provided with two spring-loaded D-shaped discs mounted on a rib across the valve bore. This design reduces the length of the path along which the center of gravity of the disc travels; it also reduces the weight of such a disc by about 50%, compared with single disc swing check valves of the same size. Coupled with spring loading, the response of the valve to retarding flow is therefore very fast.

Tilting-Disc Check Valves

Tilting-disc check valves such as the one shown in Figure 4-9 are provided with a disc-like closure member which rotates about a pivot point between the center and edge of the disc, and which is offset from the plane of the seat. The disc drops thereby into the seat on closing, and lifts out of the seat on opening. Because the center of gravity of the disc halves describes only a short path between the fully open and the closed positions, tilting-disc check valves are potentially fast closing. This particular valve is, in addition, spring loaded to ensure quick response to retarding forward flow.

Reference may be made also to the valve shown in Figure 3-70 which can serve as a butterfly valve, a tilting-disc check valve, or a combined tilting-disc check and stop valve, depending on the design of the drive.

Tilting-disc check valves have the disadvantage of being more expensive and also more difficult to repair than swing check valves. The use of tilting-disc check valves is therefore normally restricted to applications which cannot be met by swing check valves.

Diaphragm Check Valves

Diaphragm check valves such as those shown in Figures 4-10 through 4-11b are not as well known as other check valves, but they deserve attention.

The check valve shown in Figure 4-10 consists of a perforated cone-shaped basket which supports a matching diaphragm. This assembly is mounted in the pipeline between two flanges or clamped between pipe unions. Flow passing through the cone lifts the diaphragm off its seat and lets the fluid pass. When forward flow ceases, the diaphragm regains its original shape and closure is fast. One application worth mentioning is in purge-gas lines which feed into lines handling slurry or gluey substances. Under these conditions, diaphragm valves tend to operate with great reliability, while other valves hang up very quickly.

The check valve shown in Figure 4-11 uses a closure member in the form of a pleated annular rubber diaphragm. When the valve is closed, a lip of the diaphragm rests with the pleats closed against a core in the flow passage. Forward flow opens the pleats, and the lip retracts from the seat. Because the diaphragm is elastically strained in the open position, and travel of the lip from the fully open to the closed position is short, the diaphragm check valve closes extremely fast. This valve is well suited for applications in which the flow varies within wide limits. However, the pressure differential for which the valve may be used is limited to 1 MPa (145 lb/in^2), and the operating temperature is limited to about 70°C (158°F).

The closure member of the diaphragm check valve shown in Figure 4-11a consists of a flexible sleeve that is flattened at one end. The flattened end of the sleeve opens on forward flow but closes against reverse flow.

The sleeve is made in a large variety of elastomers, and is externally reinforced with plies of nylon fabric similar in construction to an automobile tire. The inside of the sleeve is soft and capable of embedding trapped solids. The valve is therefore particularly suitable for services in which the fluid carries solids in suspension or consists of a slurry.

Figure 4-11b shows an interesting application of this check valve as a tidal gate.

The valve is available in sizes as small as DN 3 (NPS 1/8) and as large as DN 3000 (NPS 120) for tidal gates.

Dashpots

The purpose of dashpots is to dampen the movement of the closure member.

The most important application of dashpots is in systems in which flow reverses very fast. If the check valve is unable to close fast enough to prevent a substantial reverse flow buildup before sudden closure, a dashpot designed to come into play during the last closing movements can considerably reduce the formation of surge pressure.

Selection of Check Valves

Most check valves are selected qualitatively by comparing the required closing speed with the closing characteristic of the valve. This selection

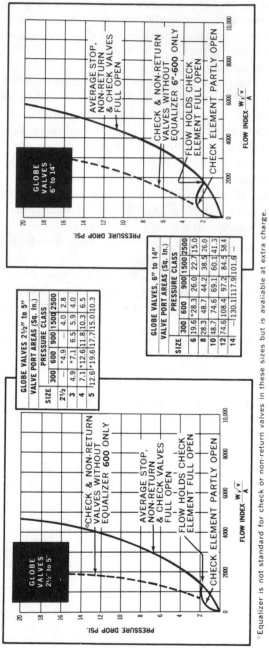

Figure 4-12. Sizing data for check valve. (Courtesy of Flow Control Division, Rockwell International.)

*Equalizer is not standard for check or non-return valves in these sizes but is available at extra charge.

CURVES SHOWING PRESSURE DROP VS. FLOW INDEX

GLOBE VALVES 2½″ to 5″

	VALVE PORT AREAS (Sq. In.)				
	PRESSURE CLASS				
SIZE	300	600	900	1500	2500
2½	—	*4.9	—	4.0	2.8
3	4.9	*7.1	6.5	5.9	4.0
4	7.1	*12.6	11.8	10.3	6.5
5	12.6	*19.6	17.7	15.0	10.3

GLOBE VALVES, 6″ to 14″

	VALVE PORT AREAS (Sq. In.)				
	PRESSURE CLASS				
SIZE	300	600	900	1500	2500
6	19.6	*28.3	26.0	22.7	15.0
8	28.3	48.7	44.2	38.5	26.0
10	48.7	74.6	69.1	60.1	41.3
12	74.6	108.4	97.2	84.5	58.4
14	—	130.1	117.9	101.6	—

GLOBE VALVES 6″ to 14″

PRESSURE DROP PSI.

AVERAGE STOP, NON-RETURN & CHECK VALVES FULL OPEN

CHECK & NON-RETURN VALVES WITHOUT EQUALIZER 6″-600 ONLY

FLOW HOLDS CHECK ELEMENT FULL OPEN

CHECK ELEMENT PARTLY OPEN

FLOW INDEX $W\sqrt{\frac{v}{A}}$

GLOBE VALVES 2½″ to 5″

PRESSURE DROP PSI.

*CHECK & NON-RETURN VALVES WITHOUT EQUALIZER 600 ONLY

AVERAGE STOP, NON-RETURN & CHECK VALVES FULL OPEN

FLOW HOLDS CHECK ELEMENT FULL OPEN

CHECK ELEMENT PARTLY OPEN

FLOW INDEX $W\sqrt{\frac{v}{A}}$

method leads to good results in the majority of applications. However, sizing is also a critical component of valve selection, as discussed in the following. If the application is critical, a reputable manufacturer should be consulted.

Check Valves for Incompressible Fluids

These are selected primarily for the ability to close without introducing an unacceptably high surge pressure due to the sudden shut-off of reverse flow. Selecting these for a low pressure drop across the valve is normally only a secondary consideration.

The first step is qualitative assessment of the required closing speed for the check valve. Examples of how to assess the required closing speed in pumping installations are given in Chapter 2, page 37.

The second step is the selection of the type of check valve likely to meet the required closing speed, as deduced from page 145.

Check Valves for Compressible Fluids

Check valves for compressible fluids may be selected on a basis similar to that described for incompressible fluids. However, valve-flutter can be a problem for high lift check valves in gas service, and the addition of a dashpot may be required.

In general, compressor-type check valves such as that shown in Figure 4-6 are the best choice for use in gas service.

Sizing of Check Valves

Check valves should be sized so that normal flow holds the closure member firmly against a stop. To permit sizing for these conditions, manufacturers must supply the appropriate sizing data. One example of such sizing data is given in Figure 4-12, in which the pressure drop is shown against a flow index, with the fully open valve position marked on the flow index curve. The flow index is given here as $W\sqrt{v}/A$, in which W = flow rate in lb/hr; v = specific volume in ft^3/lb; and A = port area in in^2. Figure 4-12 also contains tables giving the port area for a particular valve size. The valve size at which the valve is fully open at a given flow rate can thus be found.

Standards Pertaining to Check Valves

A list of USA and British standards pertaining to check valves may be found in Appendix C.

5

Pressure Relief Valves

Principal Types of Pressure Relief Valves

Pressure relief valves are pressure-relieving devices which automatically relieve a pressure system of excess pressure when abnormal operating conditions cause the pressure to exceed a set limit, and reclose when the abnormal pressure recedes again below the set limit.

Such valves may be actuated directly by the system pressure upstream of the valve, in which case the fluid load acting on the underside of the disc is opposed by a mechanical load as provided by a spring or a weight. They can also be piloted by a mechanism which senses the system pressure and pilots the pressure relief valve open or closed by removing or introducing a closing force. These modes of actuation divide pressure relief valves into direct-acting and piloted pressure relief valves.

Pressure relief valves can modulate open and closed over the entire or a substantial portion of the lift, or modulate open over only a small portion of the lift and then open suddenly to the fully open position. Accordingly, pressure relief valves may also be divided into modulating and full-lift pressure relief valves. Pressure relief valves which, by their opening characteristic, do not qualify as modulating or full-lift pressure relief valves may be classified as ordinary pressure relief valves.

The construction, application, and sizing of pressure relief valves is subject to constraints by codes, or agreement by statutory authorities. The constraints and also the definitions may vary between different codes. When applying pressure relief valves, the requirements of the applicable code must be followed or an exemption obtained from the statutory authority.

The following section discusses pressure relief valves outside the boundaries of a particular code, with the intention of giving a broader view of the subject.

Terminology

Pressure Relief Valves

Pressure relief valve: A generic term for a pressure-relieving device which automatically relieves a pressure system of excess pressure when abnormal operating conditions cause the pressure to exceed a set limit, and recloses when the abnormal pressure recedes again below the set limit.

Modulating pressure relief valve: A pressure relief valve which modulates open and closed over the entire or a substantial portion of the valve lift.

Note: For a pressure relief valve to qualify as a modulating pressure relief valve, codes may specify the minimum portion of the lift over which the valve modulates open and closed, e.g., the code in Reference 53.

Full-lift pressure relief valve: A pressure relief valve which modulates open over only a small portion of the lift and then opens suddenly to the fully open position.

Note: For a pressure relief valve to qualify as a full-lift pressure relief valve, codes may specify the maximum portion of the lift over which the valve modulates open, and the maximum overpressure prior to the valve opening suddenly, e.g., the code in Reference 53.

Ordinary pressure relief valve: A pressure relief valve with an opening characteristic which does not meet the code requirements of either the modulating or full-lift pressure relief valve.

Direct-acting pressure relief valve: A mechanically loaded pressure relief valve which is actuated directly by the pressure on the underside of the disc. Depending on operating conditions and the type of fluid for which the pressure relief valve is designed, such valves are defined as relief valves, safety valves, or safety relief valves.

Relief valve: A direct-acting pressure relief valve which is intended for liquid service only. Depending on design, such valves may open as modulating or full-lift pressure relief valves or, depending on code interpretation, as ordinary pressure relief valves.

Safety valve: A full-lift pressure relief valve which is intended for gas service only. The main application of this valve is for steam service.

Safety relief valve: A direct-acting pressure relief valve intended for gas, vapor, and liquid service. The valve opens in gas or vapor service as a full-lift pressure relief valve. In liquid service, the valve may likewise open as a full-lift pressure relief valve or, depending on code interpretation, as a modulating or ordinary pressure relief valve. Depending on construction, safety relief valves are referred to as conventional or balanced safety relief valves.

Conventional safety relief valve: A safety relief valve in which the bonnet is vented to the discharge side of the valve.

Balanced safety relief valve: A safety relief valve which is designed so that the opening and closing forces due to back pressure acting on the seated disc are balanced.

Piloted pressure relief valve: A pressure relief valve which consists of a main valve and a pilot mechanism. The main valve is the actual pressure relief valve, while the pilot mechanism senses the pressure of the pressure system and pilots the main valve open and closed.

The force to hold the main valve closed may be derived thereby from the system fluid, an external source, or a combination of these.

Piloted pressure relief valve with restricted loading: A piloted pressure relief valve in which the valve loading is restricted to permit the valve to open fully within the permissible overpressure should the pilot mechanism fail to remove the valve loading.

Piloted pressure relief valve with unrestricted loading: A piloted pressure relief valve in which the magnitude of the valve loading is unrestricted.

Piloted pressure relief valve with combined safety and control function: A piloted pressure relief valve which acts at normal operating pressures as a control valve, and at abnormal pressure as a pressure relief valve.

Terms Relating to the Actuation of Piloted Pressure Relief Valves

Energize-to-trip principle: The pilot mechanism trips the main valve open on being energized.

Deenergize-to-trip principle: The pilot mechanism trips the main valve open on being deenergized.

Energize-to-open principle: The main valve opens on being energized.

Deenergized-to-open principle: The main valve opens on being deenergized.

Pressure Terms Relating to Pressure Systems

Operating pressure: The pressure in the pressure system at normal operating conditions.

Maximum allowable working pressure: The maximum pressure at which the pressure system is permitted to operate under service conditions in compliance with the applicable construction code for the pressure system.

Accumulation: The pressure increase in the pressure system over the maximum allowable working pressure, expressed in pressure units or percent of the maximum allowable working pressure.

Maximum allowable accumulation: The maximum allowable pressure increase in the pressure system over the maximum allowable working pressure, expressed in pressure units or percent of the maximum allowable working pressure.

Pressure Terms Relating to Pressure Relief Valves

Start-to-open pressure: The pressure at which the valve commences to open.

Set pressure: The start-to-open pressure at service conditions of back pressure and temperature.
Note: This definition complies with ISO 4126.1979 (E). In USA practice, this definition applies to modulating pressure relief valves only. In the case of full-lift pressure relief valves, the pressure at which the pressure relief valve opens suddenly is considered to be the set pressure.

Cold differential test pressure: The start-to-open pressure at the test stand. This pressure includes corrections for service conditions of back pressure and/or temperature.
Note: This definition complies with ISO 4126.1979 (E). In USA practice, this definition applies to modulating pressure relief valves only. In the case of full-lift pressure relief valves, the cold differential test pressure is the pressure at which the valve opens suddenly on the test stand.

Seating pressure difference: The difference between the operating pressure and the set pressure, expressed in pressure units or percent of set pressure.

Opening pressure: The system pressure at which a full-lift pressure relief valve opens suddenly.

Opening pressure difference: The difference between the set pressure and the opening pressure, expressed in pressure units or percent of set pressure.

Overpressure: The pressure increase over the set pressure, expressed in pressure units or percent of set pressure.

Relieving pressure: Set pressure plus overpressure.

Reseating pressure: The pressure at which the pressure relief valve reseats after discharge.

Operating pressure difference: The difference between relieving pressure and reseating pressure, expressed in pressure units or percent of set pressure.

Blowdown: The pressure difference between set pressure and reseating pressure, expressed in pressure units or percent of set pressure.

Back pressure: The pressure at the outlet of pressure relief valves, expressed in pressure units or percent of set pressure.

Built-up back pressure: The pressure at the outlet of the pressure relief valve which is caused by the flow from that particular valve into the discharge system, expressed in pressure units or percent of set pressure.

Superimposed back pressure: The static pressure at the outlet of the pressure relief valve from foreign sources prior to the time the valve is required to operate, expressed in pressure units or percent of set pressure.

Miscellaneous Terms

Flow area: The minimum net area which determines the flow through the valve.

Nominal flow area: The nominal or computed flow area for use in recognized flow formulas in conjunction with the valve's certified coefficient of discharge.

Flow diameter: The diameter corresponding to the flow area.

Nominal flow diameter: The diameter corresponding to the nominal flow area.

Curtain area: The area of the cylindrical or conical discharge opening created between the seating surfaces by the lift of the disc above the seat.

Huddling chamber: The annular pressure chamber located beyond the seat of full-lift pressure relief valves.

Secondary orifice: The ring-shaped opening at the exit of the huddling chamber.

Coefficient of discharge: The ratio of the measured relieving capacity to the theoretical relieving capacity, related to the nominal flow area.

Applied coefficient of discharge: The coefficient of discharge multiplied by a derating factor, as required by codes for computing the relieving capacity.

Chatter: Rapid reciprocating motions of the disc during which the disc contacts the seat.

Flutter: Rapid reciprocating motions of the disc during which the disc does not contact the seat.

Crawl: The gradual adjustment of the set pressure of spring-loaded pressure relief valves from below normal to normal after the temperature of the spring has been raised by the fluid just discharged.

Direct-Acting Pressure Relief Valves

Development of Direct-Acting Pressure Relief Valves

The simplest and probably the earliest form of a direct-acting pressure relief valve consists of a mechanically loaded disc which covers an opening of a pressure system. When the fluid load on the underside of the disc overcomes the mechanical load on the disc, the valve opens and relieves excess pressure. However, if the shape of the disc does not permit the escaping fluid to impart much of its kinetic energy to the disc, the lift of the disc will be very small within the normally permissible overpressure. Early efforts were therefore directed to improving the valve lift.

The first significant improvement in valve lift was achieved in 1848 by an Englishman, Charles Ritchie, with a valve for the relief of gas, which utilized the expansive property of the gas for raising the disc. This was achieved by providing the disc with a peripheral lip which, in conjunction with the seat, formed an annular chamber around the seat, as shown in Figure 5-1. When a valve thus designed begins to open, the discharging gas expands in the annular chamber, but cannot readily escape. The static pressure in the annular chamber therefore rises sharply and, as it acts on an enlarged area on the underside of the disc, forces the disc open suddenly. The escaping gas deflects on the lip of the disc at 90°, thus a portion of its kinetic energy is converted into a lifting force. However, the lifting force achieved is not high enough to raise the disc to the fully open position within the normally permissible overpressure. When the disc descends again towards the seat as the overpressure recedes, the static pressure in the annular chamber builds up again. This causes the disc initially to float or huddle above the seat until the pressure has dropped below the set pressure, resulting in a pressure difference between the set pressure and the reseating pressure, known as blowdown. It is from this huddling action that the annular chamber owes the name huddling chamber. The valve may be used also for the relief of liquids. However, liquids, being incompressible, cannot develop a sudden pressure rise in the annular chamber, and the valve will open initially very little with rising overpressure.

In 1863, William Naylor introduced another improved lift pressure relief valve in which a lip around the disc turned the fluid through 180° (see Figure 5-2). This permitted the flowing fluid to impart the maximum lifting force on the disc from its momentum. However, the valve did not contain the annular chamber around the seat which causes the sudden lift at low overpressure in gas service. The lift thus achieved was not sufficient to raise the disc to the fully open position within the normally accepted overpressure.

The design of modern full-lift pressure relief valves is based on the combined principles of the Ritchie and Naylor valves, namely the expansion

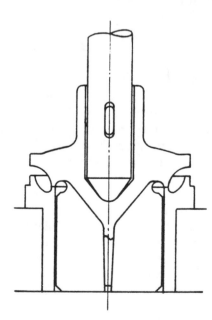

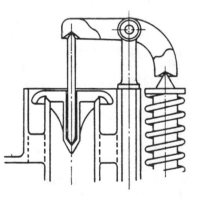

Figure 5-1. Valve elements of pressure relief valve, Ritchie type.

Figure 5-2. Pressure relief valve, Naylor type.

of a compressible fluid in an annular chamber around the seat, and the deflection of the discharging fluid through 180°.

Valve Loading

The mechanical load of direct-acting pressure relief valves may be provided either by a spring or a weight. All early pressure relief valves employed weight loading for two reasons. First, it was difficult at that time to produce a satisfactory spring for pressure relief valves. Second, objections were raised against the characteristic of the spring to raise the disc loading as the disc rises.

But on the other hand, a weight soon becomes very heavy and eventually impractical as valve size and pressure increase. Lever mounting the weight extends the range of application, but the lever weight cannot conveniently be enclosed to prevent unauthorized interference with the valve pressure setting, and it is also sensitive to vibrations, which causes unstable seat loading. The application of weight-loaded pressure relief valves is therefore limited in practice to low pressures only.

The spring has the advantage of being light-weight, and thus facilitates the construction of high-capacity and high-pressure valves. The spring can also conveniently be protected against accidental damage or unauthorized interference with its setting, and the loading from the spring is less sensitive to vibrations. The lifting-force characteristic of full-lift pressure relief valves also matches, more or less, the spring characteristic; thus the early objections to the use of springs are irrelevant now. For these reasons, most modern direct-acting pressure relief valves are spring loaded.

Modern Direct-Acting Pressure Relief Valves

All modern direct-acting pressure relief valves for high discharge capacities are designed on the combined principles of the Ritchie and Naylor valves. To avoid obstructing the flow passage which would reduce the flow rate, the disc of these valves is guided above the seat instead of in the seat bore. Two principal designs for guiding the disc have evolved.

In one design, the disc is carried by a piston traveling in a cylinder above the seat, as in the valves shown in Figures 5-3 through 5-5, 5-8, 5-9, 5-15, and 5-16. The guiding surfaces of the piston serve also as a deflector for the discharging fluid. One advantage of this guiding principle is that it permits the incorporation of effective blowdown control devices, as described later in this chapter, pages 169 and 181. This type of guiding construction is widely used in valves for the process industry, and dominates in valves for steam power plants.

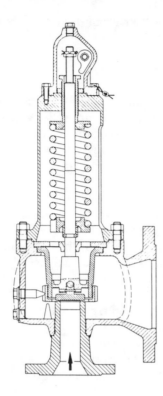

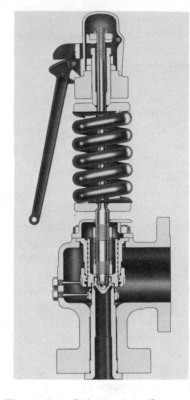

Figure 5-3. Safety valve. (Courtesy of Sempell Armaturen.)

Figure 5-4. Safety valve. (Courtesy of Dresser Industries.)

In a second design, the disc is mounted in a bell-shaped flow deflector which is guided at the back in a small-diameter cylinder below the bonnet, as in the valves shown in Figures 5-6, 5-7, and 5-10 through 5-12. One advantage of this design is that the guiding surface for the disc assembly can be completely isolated from the flowing fluid by bellows. On the debit side, sensitive blowdown control is difficult to achieve, as described on pages 169 and 181. This type of guiding construction is widely used in valves for the process industry, but seldom used in valves for steam power plants.

Fluids such as steam or dangerous or unpleasant gases and liquids demand specific valve designs. This led to the development of safety valves, safety relief valves, and relief valves.

Safety valves. These are full-lift pressure relief valves which are primarily intended for steam service. Examples of these valves are shown in Figures 5-3 through 5-5, and Figure 5-16.

One requirement of pressure relief valves in steam service is to protect the spring from excessive heat. Any rise in temperature of the spring as a result of valve discharge reduces the strength of the spring and therefore lowers the set pressure of the valve until the spring has regained its normal operating temperature. For this reason, the bonnet of these valves is normally provided with large windows which permit the spring to be cooled by the surrounding air. However, the bonnet is not normally fluid tight to the space below the bonnet, so some of the discharging steam will escape through the open bonnet. For high-temperature steam service, the bonnet is frequently separated from the valve body by a lantern-type spool piece which is designed to shield the spring from direct contact with the discharging steam, as in the valve shown in Figure 5-5.

Safety valves for steam power plants are also frequently required to permit close blowdown control. Because valve constructions in which the

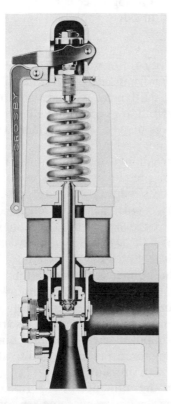

Figure 5-5. Safety valve with cooling spool. (Courtesy of Crosby Valve & Gage Company.)

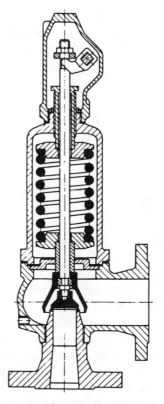

Figure 5-6. Conventional safety relief valve. (Courtesy of Bopp & Reuther GmbH.)

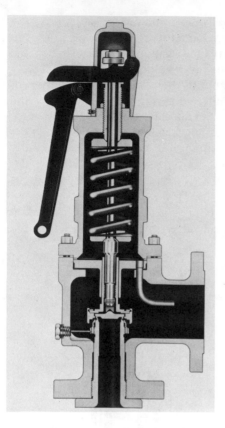

Figure 5-7. Conventional safety relief valve. (Courtesy of Dresser Industries.)

disc travels in a cylinder permit the incorporation of effective blowdown control devices, most safety valves are of this construction.

Safety valves may be used also for gases or vapors other than steam, provided they are permitted to escape into the surroundings of the valve. Generally, however, safety valves are used mainly for steam and air.

Codes commonly require that safety valves have a lifting lever by which the free mobility of the disc can be tested in service. For this reason, a lifting lever is normally standard in safety valves.

Safety relief valves. These valves differ from safety valves mainly in that the bonnet is fluid tight to the surroundings of the valve. Such valves may be used for fluids which are not permitted to escape into the surroundings of the valve. Thus, most pressure relief valves for the process industry are of this type.

Two types of safety relief valves can be distinguished: the conventional safety relief valve shown in Figures 5-6 through 5-8, and the balanced safety relief valve shown in Figures 5-9 through 5-12.

In conventional safety relief valves, the bonnet is vented to the valve outlet. This permits back pressure to act on the entire back of the disc. When the disc is seated, the area of the disc exposed to back pressure is bigger on the back than on the underside by an area equal to the seat area. This area is referred to as uncompensated area. Superimposed back pressure acting on the uncompensated area introduces a closing force on the seated disc which raises the set pressure by the amount of back pressure.

Balanced safety relief valves overcome the influence of superimposed back pressure on the set pressure by having the uncompensated area of the disc exposed to atmospheric pressure. This is commonly achieved by fitting bellows between the disc and the bonnet, whereby the bonnet is vented via a hole to the atmosphere. This vent hole may have to be piped to a safe location should the bellows fail in service. The valve shown in Figure 5-12 has, in addition, an auxiliary balanced piston which comes into play in case the bellows fail.

The behavior of safety relief valves in gas or vapor service is similar to that of safety valves with open bonnet, except in back-pressure applications. Because the control of blowdown for pressure relief valves in the process industries is frequently not important, the facilities for blowdown control in safety relief valves are normally less elaborate than in safety valves, and they are sometimes missing altogether.

Safety relief valves may also be used in liquid service, but the valve lift at 10% overpressure is relatively small.

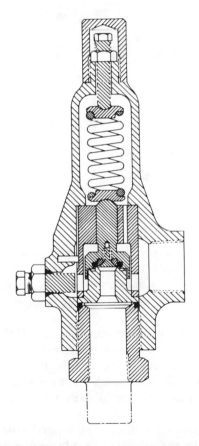

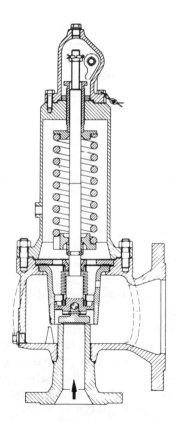

Figure 5-8. Conventional safety relief valve. (Courtesy of Anderson, Greenwood & Co.)

Figure 5-9. Balanced Safety relief valve. (Courtesy of Sempell Armaturen.)

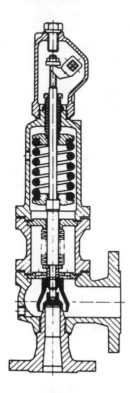

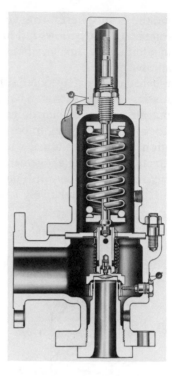

Figure 5-10. Balanced safety relief valve. (Courtesy of Bopp & Reuther GmbH.)

Figure 5-11. Balanced safety relief valve. (Courtesy of GEC-Elliot Control Valves, Limited.)

valve initially modulates open until the overpressure has risen to about 15%-20%. Then the valve opens suddenly to the fully open position, resulting in a blowdown of about 20%. These values of overpressure and blowdown may vary considerably between different makes of valves and valve sizes.

The lift prior to sudden valve opening is thereby relatively small. If the fluid pressure fluctuates — as in cases in which the system is pressurized by a reciprocating pump, or due to flow disturbances — the valve will tend to flutter or chatter in the partially open position. Also, sudden full opening of the valve produces a surge pressure which must be considered when designing the pressure system.

Codes do not normally require that safety relief valves be provided with a lifting lever. These are therefore normally special. The lifting lever may be either of the packed type, in which case the entry of the lever shaft into the bonnet is sealed by means of a packing, or of the open type, in which case the entry of the lever shaft into the bonnet is left open to the atmosphere. This construction is not permissible for conventional safety relief valves if the fluid discharged by the valve is not allowed to escape into the surroundings of the valve.

Safety relief valves, like other pressure relief valves, may be provided with soft seatings such as those shown in Figure 5-13. Soft seatings have the advantage over metal seats in that they overcome sealing problems with hard-to-seal fluids. They also tend to remain fluid tight after repeated valve operation. However, the use of soft seatings is restricted by temperature and fluid compatibility of the soft seating material. Manufacturers' catalogues may be consulted when selecting the soft seating material.

Relief valves. These are pressure relief valves which are intended for liquid service only. Depending on design, relief valves may open as modulating or full-lift pressure relief valves or, depending on code interpretation, as ordinary pressure relief valves.

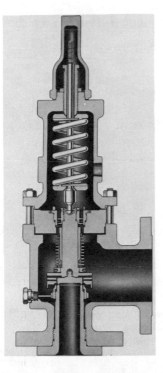

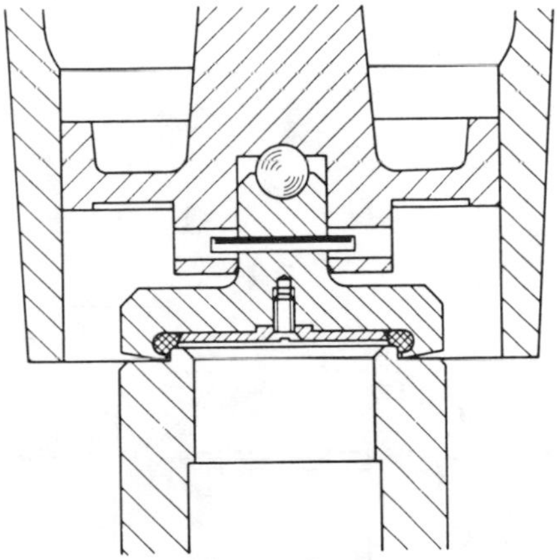

Figure 5-13. Seatings of pressure relief valve with O-ring seal. (Courtesy of Sempell Armaturen.)

Figure 5-12. Balanced safety relief valve with balanced auxiliary piston. (Courtesy of Dresser Industries.)

The simplest form of relief valve uses a plain disc without a flow deflector to increase the lifting force. Within normally accepted overpressures, these valves open therefore only by a small amount. However, such valves are still widely used where the relief of a small amount of liquid is sufficient to restore safe operating conditions; for example, in cases where the pressure rise in a fluid system is due to the thermal expansion of a liquid.

High capacity relief valves that open fully within low overpressures are of relatively recent origin. These valves differ from conventional or balanced safety relief valves only by the shape of the flow deflector around the disc, which is designed to promote a high lifting force. As a result, these valves are capable of opening fully in liquid service within an overpressure of 10%. Figures 5-14 and 5-15 show examples of such valves.

The valve shown in Figure 5-14 is capable of opening fully within an overpressure of 6%–8%. The valve may also be provided with an upper blowdown ring. In this case, the blowdown can be adjusted down to 6%–10%.

The relief valve shown in Figure 5-15 opens fully within an overpressure of about 10%, and closes from this position after a blowdown of about 20%. To combat rapid reciprocating movements of the disc, the stem of the disc moving in a guide is provided with an elastomeric friction ring which is pres-

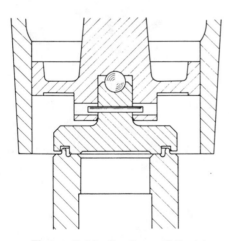

Figure 5-14. Seatings of liquid relief valve, full-lift type, fully open within 10% overpressure. (Courtesy of Sempell Armaturen.)

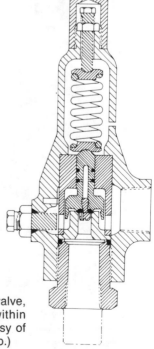

Figure 5-15. Liquid relief valve, full-lift type, fully open within 10% overpressure. (Courtesy of Anderson, Greenwood & Co.)

sure-loaded from below the disc. The resultant friction between the guide
and the moving stem tends to combat the development of rapid reciprocating
movements of the disc.

Devices for blowdown adjustment. Full-lift pressure relief valves may
be provided with devices which permit the blowdown of the valve to be
adjusted by modifying the forces acting on the disc, as explained on page
181, by means of performance diagrams.

The most common of these devices are the upper blowdown ring and the
nozzle ring.

The upper blowdown ring is used to lengthen or shorten the cylinder in
which the disc moves, as in the valves shown in Figures 5-3 through 5-5 and
5-16. Raising the blowdown ring reduces the blowdown. Conversely, lowering the blowdown ring increases the blowdown.

The nozzle ring is used to modify the width of the secondary orifice, as in
the valves shown in Figures 5-7, 5-11, and 5-12. Raising the blowdown ring
increases blowdown, but reduces at the same time the opening pressure
difference. This is not always desirable, so adjustment of the blowdown by

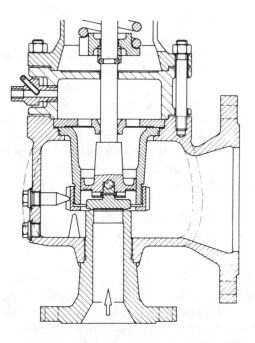

Figure 5-16. Safety valve with blowdown control throttle and upper blowdown ring.
(Courtesy of Sempell Armaturen.)

means of the nozzle ring must be carried out judiciously. Where the valve is provided with an upper blowdown ring and a nozzle ring, as in the valves shown in Figures 5-4 and 5-5, the nozzle ring is used mainly to restore the geometry around the seat after the seat has been repaired.

A third device for the adjustment of blowdown consists of a throttle which regulates the pressure build-up on the back of the disc. Such a device is shown in the valves in Figures 5-8 and 5-16. In Figure 5-16, the throttle is used in conjunction with an upper blowdown ring. This combination permits very fine blowdown control.

The need for close blowdown control exists mainly in installations where the valve blows frequently and a short blowdown is desirable for economic or other reasons.

Operation of Direct-Acting Pressure Relief Valves

Seating Pressure Difference

The seating pressure difference is the difference between the operating pressure and the set pressure. This pressure difference must be high enough to ensure sufficient seat loading for achieving a fluid-tight seat seal at normal plant operation, and higher than the blowdown to ensure that the valve closes above the operating pressure.

The minimum seating pressure difference frequently suggested for pressures up to 7 MPa (1000 lb/in^2) is 10% of the set pressure, but not less than 35 kPa (5 lb/in^2). For pressures higher than 7 MPa (1000 lb/in^2), a minimum seating pressure difference of 7% of the set pressure has been suggested.

These seating pressure differences may have to be increased for metal-seated valves if the fluid is toxic, corrosive, cryogenic, or exceptionally valuable, or if the system pressure fluctuates, as in the discharge line of reciprocating pumps and compressors.

Also, blowdown in gas or vapor service may be as high as 15%, and in liquid service as high as 20%, as discussed in the next section. If the valve is not to reseat below the operating pressure, the seating pressure difference must be accordingly high.

Blowdown

Blowdown is the pressure difference between the set pressure and the reseating pressure. From an economical point of view, the blowdown should be as short as possible to avoid unnecessary fluid loss. However, from the point of stable valve operation, a high blowdown is desirable. A compromise may therefore have to be accepted.

Codes for the application of pressure relief valves may state the limits for blowdown. The international standard ISO 4126.1979 (E) recommends the following blowdown limits for safety and safety relief valves in gas or vapor service:

1. If the blowdown is adjustable, the low blowdown limit shall be 2.5% and the high blowdown limit 7%, but with the following exceptions:
   a. If the seat bore is less than 15mm (5/8 in), the maximum blowdown should not exceed 15%.
   b. If the set pressure is less than 300 kPa (45 lb/in²), the maximum blowdown should not exceed 30 kPa (4.5 lb/in²).
2. If the blowdown cannot be adjusted, the maximum blowdown shall not exceed 15%.

The maximum blowdown recommended in this international standard for liquid service is 20%.

Opening Pressure Difference

The opening pressure difference is the pressure difference between set pressure and opening pressure of full-lift pressure relief valves. From the point of shortening the period during which the valve is just barely open, a small opening pressure difference is desirable. However, if the opening pressure difference is too small, seat leakage in gas or vapor service can be sufficient to cause the valve to pop open prematurely. Because the resultant flow would not be able to hold the valve in the open position, the valve would open and close in rapid succession. In practice, the opening pressure difference of full-lift pressure relief valves in gas or vapor service ranges between 1% and 5%. In liquid service, the opening pressure difference may vary between 6% and 20%, depending on the design of the valve.

Performance Diagram

The performance diagram displays the opening and closing forces which act on the disc* over the lift, whereby the opening force is taken at constant inlet pressure. From the relationship between the opening and closing forces over the lift, the behavior of the pressure relief valve can be explained or predicted.

Figure 5-17 shows such a performance diagram. The abscissa represents the valve lift in terms of the nominal flow diameter, and the ordinate represents the closing and opening forces in terms of these forces at the

---

*The term is used in this case to include the piston or bell-shaped disc holder on which the flowing fluid and back pressure act. Both components are frequently made of unit construction.

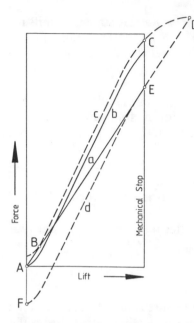

Figure 5-17. Performance diagram of safety and safety relief valves in gas or vapor service. (Courtesy of Sempell Armaturen.)

start-to-open pressure. The closing force from the spring is represented by a straight line, and the opening force from the flowing fluid by a S-shaped curve. The shape of the opening force curve depends on the type of valve, the flow medium, and the operating conditions. In this particular case, the opening force curve applies to a safety or safety relief valve in gas or vapor service, under the conditions that the pressure loss in the inlet line to the pressure relief valve and the back pressure at the outlet of the pressure relief valve are small enough to be neglected.

Opening-force curves for higher or lower inlet pressures are located parallel to each other, but at a correspondingly higher or lower level. This particular diagram shows three opening-force curves: the center one is taken at the set pressure, the top one at the opening pressure, and the bottom one at the reseating pressure.

When rising overpressure causes the valve to open from the operating point A, the closing force gains initially more than the opening force. The disc will therefore modulate open until the operating point B has been reached. Then the opening force gains more than the closing force and the disc rises sharply until arrested by a stop at the operating point C. If the disc were not arrested at C, it would travel to the operating point D where the opening and closing forces are in equilibrium. The disc might then cycle around this point. A valve of optimum design must be fully open at the operating point C.

When the inlet pressure falls again, the disc will remain in the fully open position until the operating point E has been reached. Then the opening force diminishes faster than the closing force, and the valve closes sharply to the operating point F.

The pressures at the operating points A, B, and F are the set pressure, the opening pressure, and the reseating pressure, respectively. The pressure differences between the operating points A and B, and C and F, are the opening pressure difference and the blowdown, respectively.

Performance diagrams of safety relief valves in liquid service are shown on page 179.

Influence of Back Pressure on Performance of Safety and Safety Relief Valves in Gas or Vapor Service

Back pressure introduces forces on the disc which vary in magnitude as the disc travels between the closed and open positions. These variations of force lead with increasing back pressure to unsatisfactory valve performance.

When the valve is closed, back pressure acts on both sides of the disc. The area on top of the disc exposed to back pressure differs with the design of the valve, while the exposed area on the underside of the disc is always that portion which overhangs the seat. The difference between both areas is referred to as the uncompensated area.

When the valve opens, back pressure begins to lose its influence on the underside of the disc until the velocity at the secondary orifice is sonic. At this point, back pressure has lost its influence on the underside of the disc altogether.

The following three illustrations explain the influence of back pressure on the opening and closing forces.

The first illustration, Figure 5-18a, shows the influence of built-up and superimposed back pressure on the opening force due to the back pressure acting on the underside of the disc. The various curves represent:

1. Opening force due to the flowing fluid only.
2. Opening force under the influence of built-up back pressure from flow through the valve body.
3. Opening force under the influence of built-up back pressure from flow through the valve body and discharge pipeline.
4. Opening force under the influence of superimposed back pressure.

The curves show that built-up back pressure initially increases the opening force as the valve opens. As the flow velocity at the secondary orifice approaches sonic velocity, the influence of built-up back pressure on the opening forces decreases and finally is lost completely.

Superimposed back pressure increases the opening forces already at zero valve lift. However, as the valve opens, the influence of superimposed back pressure on the opening force diminishes and eventually becomes identical with that for built-up back pressure.

The second illustration, Figure 5-18b, shows the effect of built-up back pressure on the closing forces acting on the back of the disc of conventional and balanced safety relief valves. In safety valves with open bonnet of the

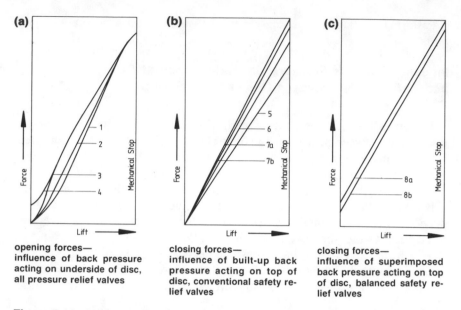

Figure 5-18. Influence of back pressure on opening and closing forces of pressure relief valves in gas or vapor service. (Courtesy of Sempell Armaturen.)

type shown in Figures 5-3 through 5-5, back pressure cannot introduce a force on the back of the disc. The various curves represent:

5. Closing force from spring only.
6. Closing force under the influence of built-up back pressure from flow through the valve body in conventional safety relief valves.
7a. Closing force under the influence of built-up back pressure from flow through the valve body and discharge pipeline in conventional safety relief valves.
7b. Closing force as under 7a, but for balanced safety relief valves.

In the case of superimposed back pressure, the closing-force component due to back pressure acting on the back of the disc is constant over the entire valve lift, thus raising the closing force curves accordingly, as shown in the third illustration, Figure 5-18c. The various curves represent:

8a. Closing force under the influence of superimposed back pressure for conventional safety relief valves.
8b. Closing force as under 8a, but for balanced safety relief valves.

The appropriate performance diagrams for back pressure can now be developed by superimposing the opening-force curves onto the closing force curves.

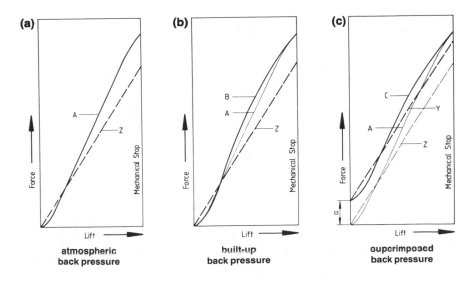

Figure 5-19. Influence of back pressure on performance of safety valves with open bonnet and piston-type disc (Figures 5-3 through 5-5). (Courtesy of Sempell Armaturen.)

In practice, the knowledge of the required spring force and its characteristic is more important than the knowledge of the sum of all closing forces. It is therefore more informative to let the closing force curve show the closing force component from the spring only, and superimpose the closing force component from the back pressure acting on the disc onto the opening force curve. The resultant force curve is referred to in the following section as the fluid-force curve, and the force as fluid force.

Performance diagram of safety valves. The performance diagrams shown in Figure 5-19 belong to safety valves of the type shown in Figures 5-3 through 5-5, which are fitted with open-bonnet and piston-type discs.

Figure 5-19a applies to the case in which the valve blows directly to the atmosphere without a discharge pipeline. The curves A and Z are repeated in Figures 5-19b and 5-19c for comparison with the curves resulting from built-up and superimposed back pressure.

In Figure 5-19b, built-up back pressure is shown to increase the fluid force only in the middle portion of the lift, resulting in the more bulgy fluid-force curve B. If the back pressure is high enough, the bulge will cause the valve initially to modulate closed from the fully open position and, as a result, raise the blowdown.

In Figure 5-19c, superimposed back pressure is shown to raise the fluid-force curve A by a margin of a, resulting in the fluid-force curve C. To regain equilibrium between fluid-force and spring force at the set pressure, the spring force Z must be increased by this margin, resulting in the spring force curve Y. If this is not done, the valve will open at a set pressure which is lower than the desired set pressure by an amount equal to 2-2½ times the superimposed back pressure. The diagram shows also that rising superimposed back pressure reduces blowdown.

Performance diagram of conventional safety relief valves. The diagrams in Figure 5-20 belong to conventional safety relief valves in which the bonnet is vented to the valve outlet, as in the valves shown in Figures 5-6 through 5-8.

Figure 5-20a applies to valves which blow directly to the atmosphere without a discharge pipeline, as represented by fluid-force curve D and the spring-force curve X. The curves A and Z from Figure 5-19a are introduced here for comparison only. The difference between the fluid-force curves A and D is brought about by the back pressure build-up in the valve housing which acts on the entire back of the disc in the case of conventional safety relief valves. To obtain the performance of the safety valves shown in Figure 5-19a, it is necessary to choose a softer spring, as shown by the spring force curve X.

The curves D and X of Figure 5-20a are repeated in Figures 5-20b and 5-20c for comparison against the curves resulting from built-up and superimposed back pressure.

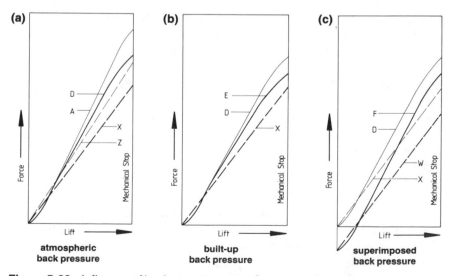

Figure 5-20. Influence of back pressure on performance of conventional safety relief valves in gas or vapor service. (Courtesy of Sempell Armaturen.)

In Figure 5-20b, built-up back pressure is shown to lower the slope of the fluid-force curve, resulting in an increase in the opening pressure difference and a reduction in blowdown, as shown by the fluid-force curve E.

In Figure 5-20c, superimposed back pressure is shown to shift the entire fluid-force curve D downward to position F. The spring-force curve must therefore be lowered to position W to regain equilibrium at the set pressure. If this is not done the valve will open at a pressure which is higher than the set pressure by an amount equal to the superimposed back pressure. Conventional safety relief valves are therefore not suitable for variable superimposed back pressure.

As the superimposed back pressure rises, the slope of the fluid-force curve F falls, resulting in an increase in the opening pressure difference and a reduction in blowdown. If the back pressure is high enough, the valve will modulate open and closed over the entire lift, resulting in a tendency of the valve to flutter or chatter.

Performance diagram of balanced safety relief valves. The diagrams in Figure 5-21 belong to balanced safety relief valves such as shown in Figures 5-9 through 5-12.

Figure 5-21a applies to valves which blow directly to the atmosphere without a discharge pipeline, as represented by the curves G and V. The curves A and Z from Figure 5-19a are introduced here for comparison only. The difference between the curves A and G is brought about by the back

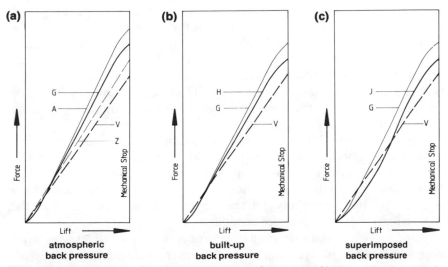

Figure 5-21. Influence of back pressure on performance of balanced safety relief valves in gas or vapor service. (Courtesy of Sempell Armaturen.)

pressure build-up in the valve housing which acts on a portion of the back of the disc in the case of balanced safety relief valves. To obtain the same valve performance as the safety valves shown in Figure 5-19a, it is necessary to choose a softer spring such as shown by the curve V. The curves G and V of Figure 5-21a are introduced in Figures 5-21b and 5-21c for comparison against the curves resulting from built-up and superimposed back pressure.

In Figure 5-21b, built-up back pressure lowers the slope of the fluid-force curve to position H, resulting in a reduction of blowdown. However, the decline of the fluid-force curve is somewhat smaller than that for conventional safety relief valves because of the smaller area on the back of the disc on which the back pressure can act.

Superimposed back pressure lowers the slope of the fluid-force curve still further, to position J, resulting in an increase in the opening pressure difference and a reduction in blowdown. If back pressure is high enough, the valve will modulate open and closed over the entire lift, resulting in a tendency of the valve to flutter or chatter.

Permissible back pressures.[61] *Safety valves* of the types shown in Figures 5-3 through 5-5 are not suitable for superimposed back pressure for two reasons: the bonnet vents to the atmosphere, and superimposed back-pressure lowers the set pressure. For this reason also, valves should not be manifolded.

However, safety valves permit a high built-up back pressure of up to 20% of the set pressure. Safety valves therefore tolerate readily the pressure drop of a diffuser-type silencer in the discharge pipeline. Also, built-up back pressure assists valve opening. On the other hand, high built-up back pressure increases blowdown, as shown in Figure 5-19b.

In *conventional safety relief valves,* the sum of built-up and superimposed back pressure is restricted to a maximum of 15% of the set pressure, allowing for an overpressure of 10%. Higher back pressures at this overpressure may reduce the valve capacity and induce the valve to flutter or chatter. The performance against back pressure can be improved by sizing the valve for a higher overpressure.

Superimposed back pressure must be constant, since variable superimposed back pressure varies the set pressure of the valve accordingly. For this reason, the exhaust pipes of conventional safety relief valves should not be manifolded. Should this be done, however, any one of the connected valves must still be able to open fully at the permissible overpressure even under the most unfavorable back-pressure conditions.

Should conventional safety relief valves be used for constant back pressure, it must be considered that the back pressure increases the set pressure by the amount of back pressure.

Balanced safety relief valves behave better against built-up back pressure than conventional safety relief valves, but identically against constant superimposed back pressure. However, superimposed back pressure has no influence on the set pressure, so balanced safety relief valves may be used for variable superimposed back pressure.

For conditions of 10% overpressure, the sum of built-up and superimposed back pressure for balanced safety relief valves should not exceed 20%. Higher back pressures are possible, but they require a softer spring for the valve. By this means, balanced safety relief valves have been made to open fully against a total back pressure of up to 50%.[61]

Influence of Back Pressure on the Performance of Safety Relief Valves in Liquid Service

The influence of back pressure on the performance of safety relief valves in liquid service is similar to the influence of back pressure in gas or vapor service. Back pressure introduces a closing force on the back of the disc in the same manner as in gas or vapor service. Two influences of back pressure are superimposed on the underside of the disc: the increase in the opening force due to back pressure acting on the underside of the disc, and the loss of mass flow due to the reduced pressure difference across the valve, which reduces the opening force. Figure 5-22 shows the resultant performance diagram for conventional safety relief valves in liquid service: Figure 5-22a for atmospheric and built-up back pressure; Figure 5-22b for superimposed back pressure.

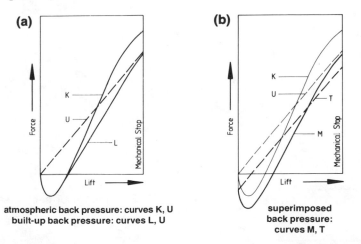

Figure 5-22. Influence of back pressure on performance of conventional safety relief valves in liquid service. (Courtesy of Sempell Armaturen.)

The curves U and K in Figure 5-22a represent the spring force and the fluid force curve at atmospheric back pressure. Built-up back pressure reduces the fluid force as the valve opens. Curve L shows the new shape of the fluid-force curve.

In the case of superimposed back pressure, the fluid force decreases over the entire lift, as represented by the curve M. The spring force must therefore be relaxed, as shown by the curve T, to regain equilibrium at the set pressure. If this is not done, the valve will open at a pressure higher than the desired set pressure by the amount of superimposed back pressure. Therefore, the valve may not be used for variable superimposed back pressure.

The performance diagram in Figure 5-23 belongs to balanced safety relief valves in liquid service. The curve S represents the spring force, and the curves N, P, and Q the corresponding fluid forces for atmospheric back pressure, built-up back pressure, and superimposed back pressure, respectively. The main difference in the performance against conventional safety relief valves is that superimposed back pressure has no influence on the set pressure of the valve.

The fluid-force curves in Figures 5-22 and 5-23 are shown to fall initially steeply before rising. As a result, the lift is initially very small with rising overpressures.

When the fluid force gains more than the spring force, the valve opens suddenly. In the case of safety relief valves operating against atmospheric back pressure, this may occur at an overpressure of about 15%-20%, resulting in a blowdown of about 20%. The lift at the point of sudden valve opening is small. For this reason, safety relief valves are not well suited for

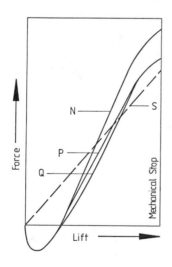

Figure 5-23. Influence of back pressure on performance of balanced safety relief valves in liquid service.

Atmospheric back pressure : curves N,S
Built-up back pressure : curves P,S
Superimposed back pressure : curves Q,S

liquid service unless they can be sized for an overpressure at which the valve opens fully.

Back pressure is shown in the performance diagrams to reduce blowdown. Thus, if back pressure is high enough, the valve will finally modulate open and closed over the entire valve lift.

The spring force/fluid force relationship that exists at atmospheric back-pressure can be restored for conditions of built-up and superimposed back pressure by using a softer spring.

Permissible Pressure Loss in Pipeline Upstream of Pressure Relief Valve for Gas or Vapor Service

The pressure loss in the pipeline upstream of the pressure relief valve reduces the mass flow by an amount directly proportional to the pressure loss. The reduced mass flow reduces the opening force from the flowing fluid and, correspondingly, the blowdown. This is shown in Figure 5-24. If the inlet pressure loss is too high, the valve will flutter. To maintain satisfactory valve operation, the pressure loss in the piping upstream of the pressure relief valve should not exceed 3% of the set pressure.

The effect of pressure loss on blowdown can be compensated for by installing a softer spring. However, this is practical only for inlet pressure losses up to 10%.[61] In sizing the pressure relief valve, the pressure loss in the inlet pipeline must be taken into account.

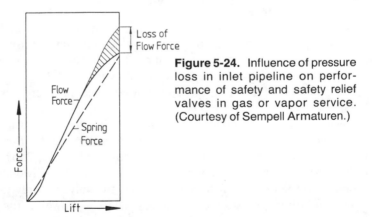

Figure 5-24. Influence of pressure loss in inlet pipeline on performance of safety and safety relief valves in gas or vapor service. (Courtesy of Sempell Armaturen.)

Adjustment of Blowdown

Various devices for the adjustment of blowdown are described on page 169. These are the upper blowdown ring found in the valves shown in Fig-

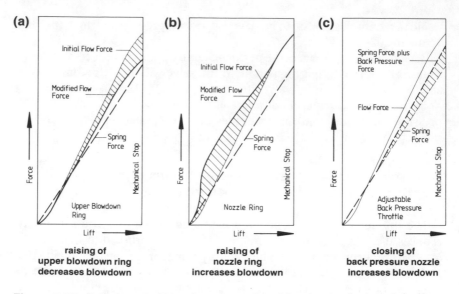

(a) raising of
upper blowdown ring
decreases blowdown

(b) raising of
nozzle ring
increases blowdown

(c) closing of
back pressure nozzle
increases blowdown

Figure 5-25. Adjustment of blowdown by various blowdown-adjustment devices. (Courtesy of Sempell Armaturen.)

ures 5-3 through 5-5, and 5-16; the nozzle ring found in the valves shown in Figure 5-7, 5-11, and 5-12; and the throttle which regulates the pressure on the back of the disc, as in the valves shown in Figures 5-8 and 5-16.

The blowdown devices achieve their purpose by modifying the flow force, as illustrated in Figure 5-25.

The upper blowdown ring adjusts the blowdown by changing the angle of deflection of the flowing fluid on the underside of the disc. Raising the upper blowdown ring decreases the angle of deflection, resulting in a decrease in the flow force and, in turn, a reduction of blowdown. This is shown in Figure 5-25a. Conversely, lowering the upper blowdown ring increases the blowdown.

The nozzle ring adjusts the blowdown by reducing or increasing the width of the secondary orifice. Raising the nozzle ring increases the flow force, but only over the middle portion of the lift, as shown in Figure 5-25b by the bulge in the flow-force curve. When the valve closes from the fully open position, the bulge in the opening-force curve causes the valve initially to modulate closed, and thus increases the blowdown. There is, however, the possibility that pressure fluctuations could cause the disc to flutter in the modulating region. Besides increasing the blowdown, raising the nozzle ring reduces the opening pressure difference. If the nozzle ring is raised too close to the disc, seat leakage may be sufficient to cause the valve to pop open prematurely. Because the flowing fluid cannot support the disc in the open position, the valve may open and close in rapid succession. The adjustment of blowdown by means of the nozzle ring must therefore be carried out judiciously.

The blowdown control throttle shown in the valves in Figures 5-8 and 5-16 is used to regulate the pressure which builds up on the back of the disc. Closing the throttle increases the pressure on the back of the disc and, in turn, the closing force. This results in a decrease in blowdown, as shown in Figure 5-25c. When using the blowdown control throttle in conjunction with an upper blowdown ring, as in the valve shown in Figure 5-16, very fine blowdown adjustment is possible.

Prime Causes of Misbehavior of Pressure Relief Valves

Pressure relief valves may behave improperly despite being of sound design and good manufacture. Such misbehavior may express itself in valve flutter or chatter, excessively high blowdown, and seat leakage. The causes of misbehavior may be wrong installation, wrong valve selection, oversizing, or non-observance of system conditions.

Valve flutter or chatter. This may be caused by the following:
1. The valve is oversized. Remedy: Restrict valve lift.
2. The valve is sized for maximum flow, but pressure upsets are only minor. Remedy: Install at least two pressure relief valves with staggered settings, with the smallest valve handling minor pressure upsets.
3. The pressure loss in the upstream system between the source of pressure and the valve inlet is too high.
4. Back pressure is too high.
5. The valve is mounted too close to equipment which causes excessive flow turbulence or pressure fluctuations. Such equipment may be a pressure reducing valve, an orifice plate, a flow nozzle, a pipe fitting such as an elbow, or a reciprocating pump or compressor. To combat valve flutter or chatter in these cases, the valve should be mounted as far away as practical from the source of flow turbulence or pressure fluctuations. If possible, the valve should be adjusted for high blowdown and/or sized for a higher overpressure.
6. The opening pressure difference is too small, so normal leakage flow is sufficient to pop the valve open prematurely.
7. Seat leakage is too high and causes the valve to pop open prematurely.

Excessively high blowdown. This may be caused by:
1. Wrong adjustment of blowdown-adjustment devices.
2. The discharging gas carries liquid droplets.
3. In the case of safety valves with an open bonnet, built-up back pressure is too high.

Valve leakage. The cause of this might be:

1. Worn seatings.
2. In the case of metal seatings, the valve operates too close to the set pressure.
3. Excessive pipeline stresses, which distort the valve body.

Piloted Pressure Relief Valves

Piloted pressure relief valves consist of a main valve which is the actual pressure relief valve, and a pilot mechanism which senses the pressure of the pressure system and pilots the main valve open and closed. The main valve consists of a valve body with a closure member and an actuator, and the actuator may be either integral with or external to the valve body. The main valve may consist also of a direct-acting pressure relief valve to which an actuator has been added. Piloted pressure relief valves may also be designed to serve at normal operating conditions as a control valve, and at abnormal operating conditions as a pressure relief valve.

The force to hold the main valve closed, and the energy to operate the main valve and pilot mechanism may be derived from the system fluid, an external source, or a combination of these. When the system pressure commences to exceed a set limit, the pilot mechanism either removes or reduces the closing force to permit the system fluid to force the main valve open, or introduces an opening force. When the system pressure recedes again, the pilot mechanism either reintroduces the closing force or removes the opening force. By this mode of opening and closing, a high closing force can be maintained just prior to valve opening. Piloted pressure relief valves are therefore capable of maintaining a high degree of seat tightness even when operating close to the set pressure.

The magnitude of the closing force just prior to valve opening may be unrestricted or, with some designs, restricted by choice. The closing force is said to be restricted if the system pressure can open the main valve within a permissible overpressure should the pilot mechanism fail to operate. The installation requirements for these valves are therefore normally more relaxed than they are for those with unrestricted loading.

Operating Media

The safest operating medium from the point of availability is the system fluid. However, the operating mechanism must be designed to tolerate the particular type of fluid at operating conditions.

Operating media of an external source, on the other hand, may be chosen to facilitate the design of piloted pressure relief valves for difficult applications. Common such operating media are compressed air, hydraulic fluids, and electric power.

Operating Principles of Main Valves

Main valves may be designed to open in response to the actuator either being energized or deenergized.

Under the energize-to-open principle, loss of power to the actuator will cause the main valve to remain closed in an emergency. The system fluid is, therefore, the safest operating medium in this case. However, some codes also permit external operating media if they come from a number of independent sources, or if the valve loading is restricted.

Under the deenergize-to-open principle, loss of power to the actuator will cause the main valve to open automatically below or at the set pressure, depending on the design of the valve. Thus the fail-safe operation of the main valve is not influenced in this case by the operating medium.

Operating Principles of Pilot Mechanisms

The pilot mechanism may be designed to trip the main valves open in response to being either energized or deenergized.

Under the energize-to-trip principle, loss of power will cause the main valve to remain closed in an emergency. The system fluid is, therefore, the safest operating medium in this case. However, some codes also permit external operating media if they come from a number of independent sources, or if the valve loading is restricted.

Under the deenergize-to-trip principle, loss of power will trip the main valve open. Thus the fail-safe operation of the pilot mechanism is not influenced in this case by the operating medium.

Main Valves for Operation by the System Fluid

The operating principles of main valves permit the design of numerous variations; the valves diagrammed in Figures 5-26 through 5-29 are typical. The first two valves operate on the deenergize-to-open principle, and the last two on the energize-to-open principle.

The closing force of these valves is provided in three ways: first, by the fluid pressure acting directly on the disc, as in the valves shown in Figures 5-26 and 5-29; second, by a spring, as in the valve shown in Figure 5-28; and third, by the fluid pressure acting on the actuator piston, as in the valve shown in Figure 5-27. The actuator piston of the latter type of valve must be provided with an effective peripheral seal capable of sealing against the full line pressure while the valve is closed.

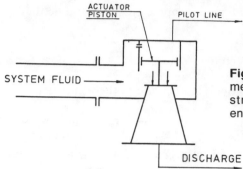

Figure 5-26. Main valve. Operating medium: system fluid. Loading: unrestricted. Operating principle: de-energize-to-open.

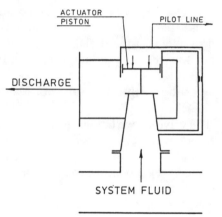

Figure 5-27. Main valve. Operating medium: system fluid. Loading: unrestricted. Operation principle: de-energize-to-open. Note: Actuator piston must seal against the full system pressure while the valve is closed.

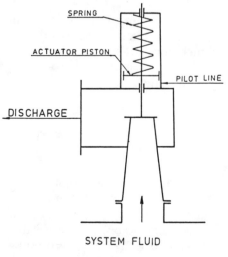

Figure 5-28. Main valve. Operating medium: system fluid. Loading: unrestricted. Operating principle: energize-to-open.

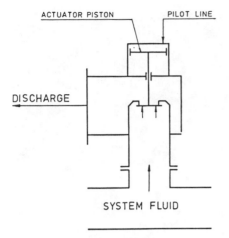

ACTUATOR PISTON   PILOT LINE

DISCHARGE

SYSTEM FLUID

Figure 5-29. Main valve. Operating medium: system fluid. Loading: restricted or unrestricted. Operating principle: energize-to-open.

Figures 5-30 through 5-34 show examples of main valves; those shown in Figures 5-30 through 5-33 operate on the deenergize-to-open principle, and the valve shown in Figure 5-34 operates on the energize-to-open principle. The valve shown in Figure 5-33 resembles the valve in Figure 5-27 in that the piston must seal against the full pressures of the system fluid while the valve is closed.

The valves shown in Figures 5-30 through 5-32 and Figure 5-34 are used in high-pressure and high-temperature steam service. Those operating on the deenergize-to-open principle and shown in Figures 5-30 through 5-32 have the advantage in steam service in that the steam forcing the valve closed holds the actuator at a temperature not lower than the saturated steam temperature. This protects the actuator against excessive thermal shock on valve opening. Any condensate in the pilot line or actuator must also be able to drain freely past the piston or through a drain hole.If this does not occur, revaporation of the condensate when venting the actuator may seriously delay the opening of the valve.

The problem of condensate revaporizing on valve opening does not exist in main valves which open on the energize-to-open principle, as in the valve shown in Figure 5-34. To combat thermal shock from the steam entering the actuator on valve opening, the actuator is arranged in the steam entry branch of the valve body where it is held externally at a temperature not lower than the saturated steam temperature.

Figures 5-35 and 5-36 show typical installation diagrams for these valves in steam power plants in compliance with the requirements of the code of practice Reference 54. The main valves operate on the deenergize-to-open and energize-to-open principles, respectively. The pilot valves, which in this

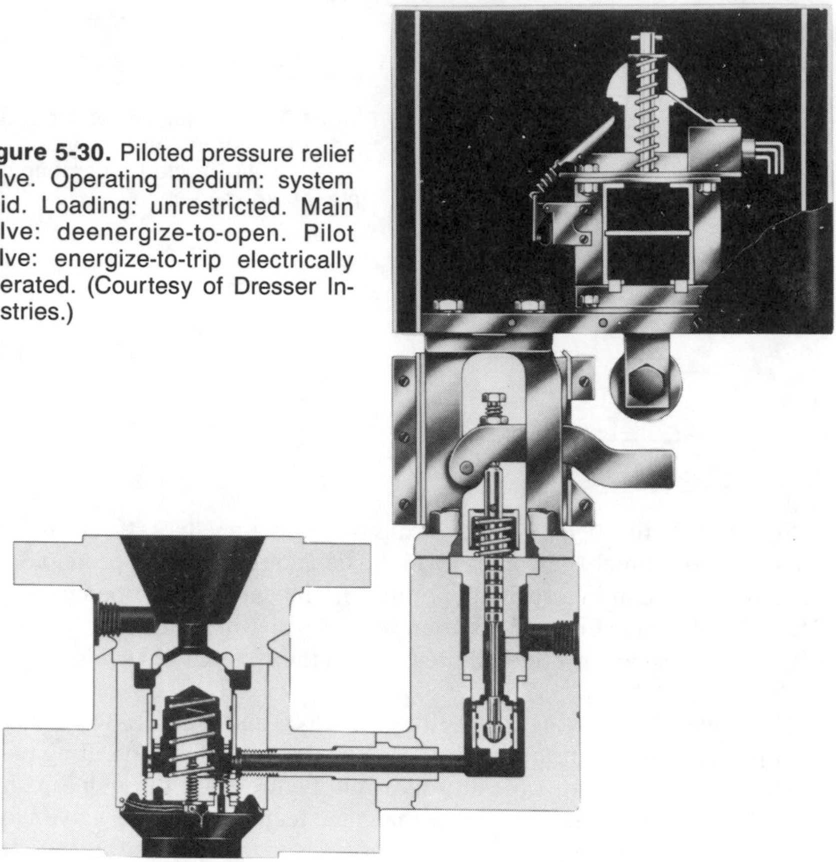

Figure 5-30. Piloted pressure relief valve. Operating medium: system fluid. Loading: unrestricted. Main valve: deenergize-to-open. Pilot valve: energize-to-trip electrically operated. (Courtesy of Dresser Industries.)

case are powered by the system fluid, operate on the energize-to-trip principle.

To achieve maximum operational reliability, the main valves are provided with three independently operating pilot valves. The number of main valves served by one set of pilot valves may thereby be greater than one. Each pilot valve may also be isolated from the main valve and tested in service for proper functioning. To prevent the main valve from becoming completely isolated from the pilot valves, the isolating valves must be mutually interlocked so that only one pilot valve can be isolated at a time. In each case, one of the pilot valves is also provided with a solenoid which permits the valve to be operated manually from a remote location.

The pilot valve, which is part of the main valve shown in Figure 5-30, operates on the energize-to-trip principle, but is powered electrically. Thus, should the electric supply to the solenoid fail, the main valve will not open in

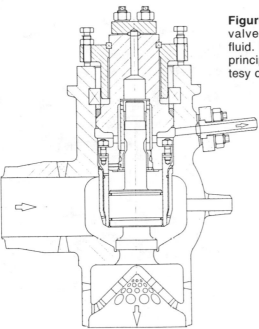

Figure 5-31. Piloted pressure relief valve. Operating medium: system fluid. Loading: unrestricted. Operating principle: deenergize-to-open. (Courtesy of Sempell Armaturen.)

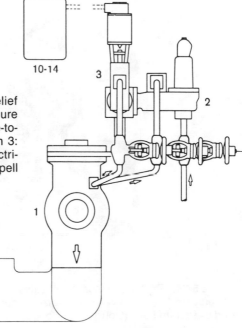

Figure 5-32. Piloted pressure relief valve. Item 1: main valve, as in Figure 5-31. Item 2: pilot valve, energize-to-trip, operated by system fluid. Item 3: pilot valve, deenergize-to-trip, electrically operated. (Courtesy of Sempell Armaturen.)

Figure 5-33. Piloted pressure relief valve. Operating medium: system fluid. Loading: unrestricted. Main valve: deenergize-to-open. Pilot valve: energize-to-trip, operated by system fluid. (Courtesy of Anderson, Greenwood & Co.)

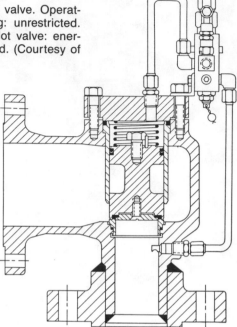

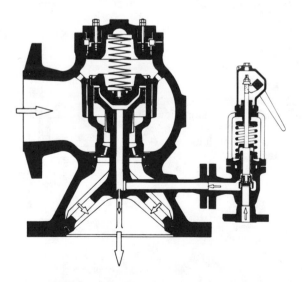

Figure 5-34. Piloted pressure relief valve. Operating medium: system fluid. Loading: unrestricted. Main valve: energize-to-open. Pilot valve: energize-to-trip, operated by system fluid. (Courtesy of Bopp & Reuther GmbH.)

1. Main valve
2. Shut-off valve arrangement, interlocked
3. Pressure indicator
4. Direct acting pressure relief valve
5. Pilot valve
6. Solonoid for remote operation of pilot valve
7. Solonoid switch
8. Adjustable throttle

a. Pressure tapping line
b. Signal line
c. Pilot line
d. Vent line

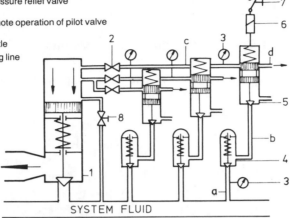

Figure 5-35. Installation diagram of piloted pressure relief valve. Operating medium: system fluid. Loading: unrestricted. Main valve: deenergize-to-open. Pilot valve: energize-to-trip. (Reprinted from German Code of Practice for Steam Boilers: Pressure Relief Valves, SR-Sicherheitsventile, 1972, courtesy of Vereinigung der Technischen Ueberwachungsvereine, e.V., Essen.)

1. Main valve
2. Shut-off valve arrangement, interlocked
3. Pressure indicator
4. Solonoid switch
5. Adjustable throttle
6. Solonoid for remote operation of pilot valve
7. Pilot valve

a. Pressure tapping line
b. Pilot line
c. Vent line

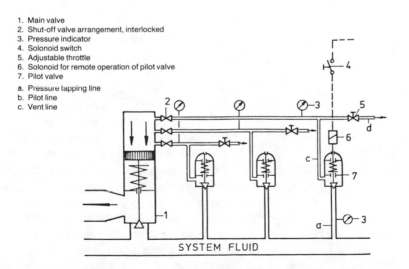

Figure 5-36. Installation diagram of piloted pressure relief valve. Operating medium: system fluid. Loading: unrestricted. Main valve: energize-to-open. Pilot valve: energize-to-trip. (Reprinted from German Code of Practice for Steam Boilers: Pressure Relief Valves, SR-Sicherheitsventile, 1972, courtesy of Vereinigung der Technischen Ueberwachungsvereine, e.V., Essen.)

an emergency. This combination of main valve and pilot valve is intended therefore to serve as a supplementary pressure relief valve only. The main valve may also be provided, of course, with a pilot valve which operates on the deenergize-to-open principle or is powered by the system fluid.

Main Valves for Operation by an External Medium

The valves shown in Figures 5-37 and 5-38 are spring-loaded main valves in which the actuator is powered by an external medium. The media used in this case are compressed air for the first valve and electricity for the second valve.

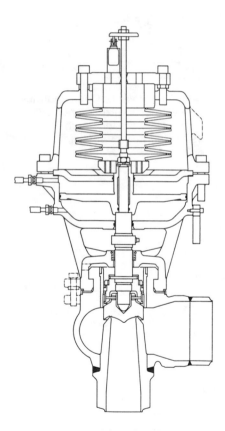

Figure 5-37. Main valve. Operating medium: external. Loading: restricted or unrestricted. Operating principle: deenergize-to-open. (Courtesy of Bopp & Reuther GmbH.)

The valve shown in Figure 5-37 operates on the deenergize-to-open principle. The actuator consists of an air motor which, at normal operating conditions, exerts a down thrust on the valve stem and, in conjunction with the valve spring, holds the valve closed. When the operating pressure reaches the set pressure, the air motor exerts a lifting force which, in conjunction with the fluid force on the underside of the disc, opens the valve. Should the power supply to the air motor fail, the valve opens as a direct-acting pressure relief valve.

The second valve, shown in Figure 5-38, operates, likewise, on the deenergize-to-open principle. When the operating pressure is normal, the solenoid acting on the stem of the valve is energized and, in conjunction with the main spring, holds the valve closed. When the operating pressure reaches the set pressure, the solenoid is deenergized and the valve opens as a direct-acting pressure relief valve.

Figures 5-39 through 5-41 show typical installation diagrams for steam power plants in compliance with the requirements of the code of

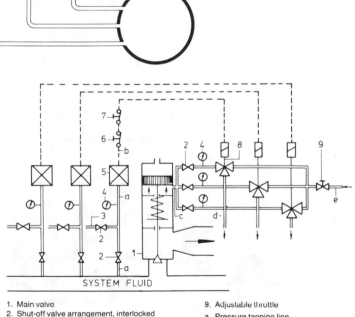

Figure 5-38. Piloted pressure relief valve. Operating medium: external. Loading: restricted or unrestricted. Main valve: deenergize-to-open. Pilot mechanism: energize-to-trip. (Courtesy of Sempell Armaturen.)

1 Pressure Switch
2 Test Connection

1. Main valve
2. Shut-off valve arrangement, interlocked
3. Test connection
4. Pressure indicator
5. Pressure switch
6. Solonoid switch (local)
7. Solonoid switch (control room)
8. Solonoid operated pilot valve
9. Adjustable throttle
a. Pressure tapping line
b. Signal line
c. Pilot line
d. Vent line
e. Supply line carrying external operating medium

Figure 5-39. Installation diagram of piloted pressure relief valve. Operating medium: external. Loading: unrestricted. Main valve: deenergize-to-open. Pilot valve: deenergize-to-trip. (Reprinted from German Code of Practice for Steam Boilers: Pressure Relief Valve, SR-Sicherheitsventile, 1972, courtesy of Vereinigung der Technischen Ueberwachungsvereine, e.V., Essen.)

194   Valve Selection Handbook

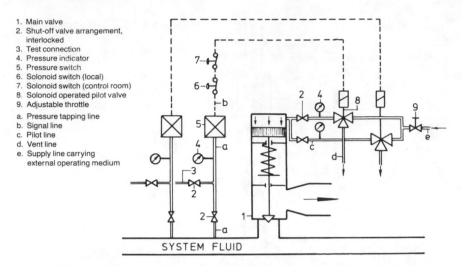

1. Main valve
2. Shut-off valve arrangement, interlocked
3. Test connection
4. Pressure indicator
5. Pressure switch
6. Solenoid switch (local)
7. Solenoid switch (control room)
8. Solenoid operated pilot valve
9. Adjustable throttle

a. Pressure tapping line
b. Signal line
c. Pilot line
d. Vent line
e. Supply line carrying external operating medium

Figure 5-40. Installation diagram of piloted pressure relief valve. Operating medium: external. Loading: restricted. Main valve: deenergize-to-open. Pilot valve: deenergize-to-trip. (Reprinted from German Code of Practice for Steam Boilers: Pressure Relief Valve, SR-Sicherheitsventile, 1972, courtesy of Vereinigung der Technischen Ueberwachungsvereine, e.V., Essen.)

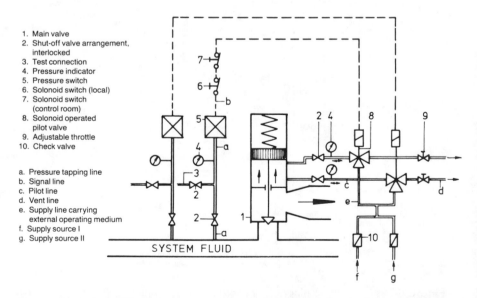

1. Main valve
2. Shut-off valve arrangement, interlocked
3. Test connection
4. Pressure indicator
5. Pressure switch
6. Solenoid switch (local)
7. Solenoid switch (control room)
8. Solenoid operated pilot valve
9. Adjustable throttle
10. Check valve

a. Pressure tapping line
b. Signal line
c. Pilot line
d. Vent line
e. Supply line carrying external operating medium
f. Supply source I
g. Supply source II

Figure 5-41. Installation diagram of piloted pressure relief valve. Operating medium: external. Loading: restricted. Main valve: energize-to-open. Pilot valve: deenergize-to-trip. (Reprinted from German Code of Practice for Steam Boilers: Pressure Relief Valve, SR-Sicherheitsventile, 1972, courtesy of Vereinigung der Technischen Ueberwachungsvereine, e.V., Essen.)

practice Reference 54. According to this code, main valves with unrestricted loading must be provided with a control mechanism in triplicate, while the number of control mechanisms for the main valve with restricted loading may be limited to two. The pilot valves operate in all these cases on the de-energize-to-trip principle.

Each pilot valve with an associated pressure switch can be isolated from the main valve and the pressure system for in-service testing. The isolated valves must be mutually interlocked so that only one pilot valve with an associated pressure switch can be isolated at a time. Provision is also made to operate one of the pilot vavles manually from a local station or the control room.

Main Valves with Combined Safety and Control Functions

The function of pressure relief valves may also be combined with that of control valves, as in the case of high-pressure bypass/safety valves shown in Figures 5-42 and 5-43 for a steam power plant. The main valve is equipped with a hydraulically powered actuator which operates on the deenergize-to-open principle, while the pilot valves in the hydraulic control lines operate on the deenergize-to-trip principle.

Figure 5-44 represents a typical installation diagram for these valves.

When the plant is in normal operation, the valve operates as an automatic high-pressure bypass valve which feeds into the reheater system. The latter is connected to the condenser via a low-pressure bypass valve, and also protected by a separate safety valve which discharges directly to the atmosphere. Should the pressure in the high-pressure system rise suddenly due to tripping of the turbine, the safety pilot valves deenergize the actuator. The steam pressure then opens the main valve rapidly, aided by a spring which acts on the actuator piston in the opening direction.

Standards Pertaining to Pressure Relief Valves

Standards pertaining to pressure relief valves, including recommended practices for the design and installation of pressure relief systems, may be found in Appendix C.

Calculation of Flow Through Pressure Relief Valves and Associated Piping

Scope

The calculations cover the sizing of pressure relief valves and diffuser-type silencers for pressure relief valves, and the determination of pressure losses in inlet and discharge pipelines and the discharge reactive force.

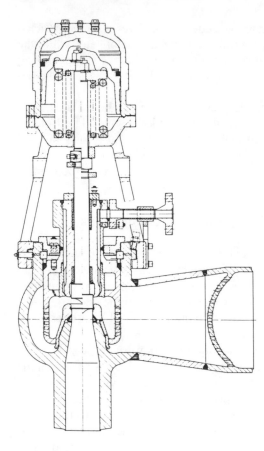

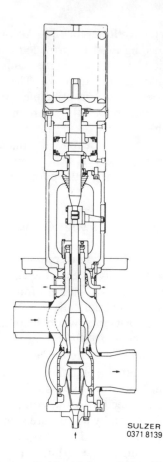

SULZER
0371 8139

Figure 5-42. Main valve with combined safety and control function. Operating medium: external. Operating principle: deenergize-to-open. (Courtesy of Bopp & Reuther GmbH.)

Figure 5-43. Main valve with combined safety and control function (see Figure 5-44 for installation diagram). Operating medium: external. Operating principle: deenergize-to-open.

Mathematical Units

The equations may be solved in coherent or mixed SI units, mixed imperial units, or any consistent set of units. The numerical values in the equations are collected in a constant B which applies to the system of units used in solving the equation. The sample calculations and the derivation of the equations in Appendix A are carried out in coherent SI units.

Nomenclature

The nomenclature is shown in Table 5-1.

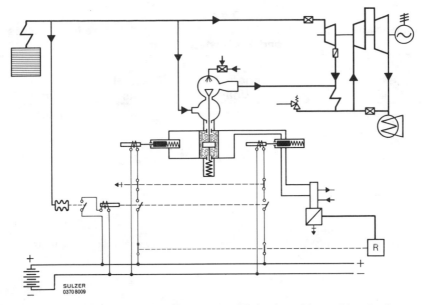

Figure 5-44. Installation of piloted pressure relief valve with combined safety and control function in steam power plant. Operating medium: external. Main valve: deenergize-to-open. Pilot valves: deenergize-to-trip. (Courtesy of Sulzer Bros. (UK) Limited.)

General Formula for Sizing of Pressure Relief Valves

The flow area of a pressure relief valve that is necessary to prevent excessive system pressure can be calculated using the following general formula:

$$A = \frac{W}{BKG} \tag{5-1}$$

Theoretical Mass Flux

There are three main categories of mass flow through pressure relief valves. Two of these have sub-categories:

1. Gas, or vapor, flow
   a. non-choked flow
   b. choked flow
2. Liquid flow
3. Mixed phase flow (vapor plus liquid)
   a. gas containing liquid droplets
   b. liquid containing gas bubbles
   c. liquid flashing to vapor through valve

Table 5-1
**Nomenclature Used in Calculating Flow Through Pressure Relief Valves
and Associated Piping**

		Coherent SI Units	Mixed SI Units	Mixed imperial Units
A	flow area	m^2	mm^2	$in.^2$
B	constant			
c	velocity of sound	m/s	m/s	ft/s
c_p	specific heat of gas at constant pressure	J/kg·K	J/kg·K	Btu/lb°F
c_v	specific heat of gas at constant volume	J/kg·K	J/kg·K	Btu/lb°F
d	internal pipe diameter	m	mm	in.
f	pipe friction factor = number of velocity heads lost in length of pipe equal to diameter			
F	discharge reactive force	N	N	$lb/in.^2$
g	local acceleration due to gravity	$9.807 \ m/s^2$	$9.807 \ m/s^2$	$32.174 \ ft/s^2$
G	mass flux	kg/m^2s	kg/mm^2h	$lb/in.^2h$
H	enthalpy of gas	J/kg	J/kg	Btu/lb
k	isentropic coefficient			
K	applied coefficient of discharge of pressure relief valve			
K_b	capacity correction factor due to back pressure, used in the sizing equations of safety relief valves for gas or vapor service			

Table 5-1
Continued

	Coherent SI Units	Mixed SI Units	Mixed imperial Units
K_d coefficient of discharge of pressure relief valves, expressing the ratio of the measured relieving capacity to the theoretical relieving capacity, based on the nominal flow area			
K_n Napier capacity correction factor for pressure			
K_{SH} Napier capacity correction factor for superheat			
K_v capacity correction factor due to viscosity, used in the sizing equations of pressure relief valves for liquid service			
K_z coefficient of discharge of nozzles in diffuser-type silencer			
I length of pipeline	m	mm	in.
M molecular weight			
P pressure (absolute)	Pa	MPa	lb/in.^{2}
P_c critical pressure (absolute), the maximum pressure at which the vapor and liquid phases of a liquid exist in equilibrium	Pa	MPa	lb/in.^{2}

Table 5-1
Continued

		Coherent SI Units		Mixed SI Units		Mixed imperial Units	
P_r	reduced pressure at valve inlet, P_1/P_c						
Q	volume rate of flow	m^3/s		m^3/min		ft^3/min	
Q_h	heat flow rate	W		W		btu/h	
r	radius	m		mm		in.	
R	gas constant	$\dfrac{8314.3}{M}$	$\dfrac{J}{kg \cdot K}$	$\dfrac{8314.3}{M}$	$\dfrac{J}{kg \cdot K}$	$\dfrac{1545}{M}$	$\dfrac{ft \cdot lbf}{lb \cdot {}^\circ R}$
T	temperature (absolute)	$K = {}^\circ C + 273$		$K = {}^\circ C + 273$		${}^\circ R = {}^\circ F + 460$	
T_c	critical temperature (absolute), the maximum temperature at which the vapor and liquid phases exist in equilibrium.	$K = {}^\circ C + 273$		$K = {}^\circ C + 273$		${}^\circ R = {}^\circ F + 460$	
T_r	reduced temperature at valve inlet, T_1/T_c						
v	flow velocity	m/s		m/s		ft/s	
V	specific volume	m^3/kg		m^3/kg		ft^3/lb	
W	mass rate of flow	kg/s		kg/h		lb/h	
Z	compressibility factor for the deviation of a real gas from the ideal gas, P_1V_1/RT_1						
Δ	differential between two points						
ρ	density	kg/m^3		kg/m^3		lb/ft^3	
σ	open area ratio of diffuser in diffuser-type silencer						
ζ	resistance coefficient						

Subscripts:

l   liquid phase

v   vapor phase

The subscripts i, n, s, t, and 1-6 refer to sections shown in Figures 5-45 and 5-46. Subscript sn refers to conditions at outlet of diffuser nozzles at section "s" in Figure 5-46.

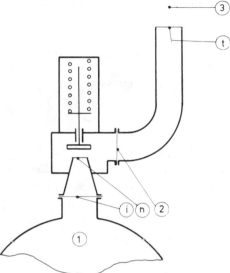

Figure 5-45. Layout of pressure relief valve system; subscripts used in flow equations.

Figure 5-46. Layout of pressure relief valve system with diffuser-type silencer; subscripts used in flow equations.

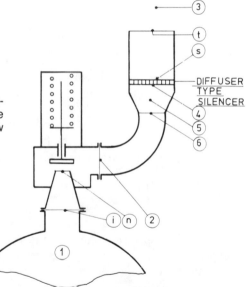

Each of these has an appropriate formula, or procedure, for determining the theoretical mass flux, G, to be used in Equation 5-1 for calculating the required flow area.

Required Mass Flow

The required mass flow through a pressure relief valve must be determined by careful examination of the effects of normal and abnormal process conditions. Some of the abnormal conditions that may cause a pressure relief valve to operate are:

1. External fire
2. Failure of reflux or cooling system
3. Closed outlet valve at a vessel, pump, or compressor
4. Vaporization in a heat exchanger
5. Heat exchanger tube failure
6. Failure of pressure-reducing control valve

Under fire conditions, or if there is a net heat input into a vessel containing vapor and liquid, the required mass flow through the pressure relief valve necessary to prevent the build-up of excessive pressure depends on whether the valve is located in the vapor space or below the liquid level.

If the valve is located in the vapor space of the vessel, the required mass flow through the valve is, as shown by Sallet:[66]

$$W = \frac{(1 - V_\ell/V_v)Q_h}{H_{v,1} - H_{\ell,1}} \tag{5-2}$$

where
Q_h = mean rate of heat flow into the vessel
$H_{v,1} - H_{\ell,1}$ = enthalpy of vaporization of fluid at stagnation condition (in vessel)

If the valve is connected to the liquid space of the vessel, Sallet has shown that the required massflow through the valve is:

$$W = \frac{(V_v/V_\ell - 1)Q_h}{H_{v,1} - H_{\ell,1}} \tag{5-3}$$

Because the specific volume of the gas phase is much greater than the liquid phase, the mass flow rate required to prevent excessive pressure is much less if the valve is located in the vapor space than if it is located in the liquid space. For this reason, pressure relief valves should be located in the vapor space rather than the liquid space of the vessel.

Sizing for Gas and Vapor

The equations for the sizing of pressure relief valves are based on the ideal gas laws in which it is assumed that flow through the valve is isentropic.

Non-choked flow. Flow through the pressure relief valve will be non-choked if the ratio of the absolute pressures at outlet and inlet is higher than critical; i.e., if:

$$\frac{P2}{P1} > \left(\frac{2}{k+1}\right)^{k/(k-1)} \qquad \text{or } P_2 > \approx 0.5P_1$$

The theoretical mass flux under these conditions is:

$$G = \left\{ \frac{P_1}{V_1} \frac{2k}{k-1} \left[\left(\frac{P_2}{P_1}\right)^{2/k} - \left(\frac{P_2}{P_1}\right)^{(k+1)/k} \right] \right\}^{1/2} \qquad (5\text{-}4)$$

For a real gas the specific volume may be found from the equation $V = RTZ/P$. Introducing this into Equation 5-4 gives an alternative theoretical mass flux equation:

$$G = P_1 \left\{ \frac{1}{RT_1 Z} \frac{2k}{k-1} \left[\left(\frac{P_2}{P_1}\right)^{2/k} - \left(\frac{P_2}{P_1}\right)^{(k+1)/k} \right] \right\}^{1/2} \qquad (5\text{-}5)$$

Thus, after substituting these equations into Equation 5-1, the nominal required flow area may be calculated from:

$$A_n = \frac{W}{BK} \left\{ \frac{P_1}{V_1} \frac{2k}{k-1} \left[\left(\frac{P_2}{P_1}\right)^{2/k} - \left(\frac{P_2}{P_1}\right)^{(k+1)/k} \right] \right\}^{-1/2} \qquad (5\text{-}6)$$

in which

B (coherent SI units)   = 1.0

$$\text{B (mixed SI units)} \qquad = \frac{3{,}600 \times (10^6)^{1/2}}{10^6} = 3.6$$

$$\text{B (mixed imperial units)} = \frac{3{,}600}{144} (32.174 \times 144)^{1/2} = 1{,}701.66$$

or

$$A_n = \frac{W}{BKP_1} \left\{ \frac{1}{RT_1 Z} \frac{2k}{k-1} \left[\left(\frac{P_2}{P_1}\right)^{2/k} - \left(\frac{P_2}{P_1}\right)^{(k+1)/k} \right] \right\}^{-1/2} \qquad (5\text{-}7)$$

in which

B (coherent SI units) $= 1.0$

B (mixed SI units) $= \dfrac{3,600 \times 10^6}{10^6} = 3,600$

B (mixed imperial units) $= 3,600(32.174)^{1/2} = 20,420$

Choked flow. Flow through the pressure relief valve will be choked if the ratio of absolute pressures at inlet and outlet is less than critical; i.e., if:

$$\frac{P2}{P1} < \left(\frac{2}{k+1}\right)^{k/(k-1)} \qquad \text{or } P_2 < \approx 0.5P_1$$

The theoretical mass flux under these conditions is:

$$G = \left[\frac{P_1}{V_1} k \left(\frac{2}{k+1}\right)^{(k+1)/(k-1)}\right]^{1/2} \tag{5-8}$$

For a real gas the specific volume may be found from $V = RTZ/P$. Introducing this into Equation 5-8 gives an alternative mass flux equation:

$$G = P_1 \left[\frac{2}{ZRT_1} \left(\frac{2}{k+1}\right)^{(k+1)/(k-1)}\right]^{1/2} \tag{5-9}$$

After substituting these equations into Equation 5-1, the nominal required flow area may be calculated from:

$$A_n = \frac{W}{BKK_b} \left[\frac{P_1}{V_1} k \left(\frac{2}{k+1}\right)^{(k+1)/(k-1)}\right]^{-1/2} \tag{5-10}$$

in which

B (coherent SI units) $= 1.0$

B (mixed SI units) $= \dfrac{3,600 \times (10^6)^{1/2}}{10^6} = 3.6$

B (mixed imperial units) $= \dfrac{3,600}{144} (32.174 \times 144)^{1/2} = 1,701.66$

or

$$A_n = \frac{W}{BKK_bP_1} \left[\frac{1}{RT_1Z} \; k \left(\frac{2}{k+1}\right)^{(k+1)/(k-1)}\right]^{-1/2}$$ (5-11)

in which

B (coherent SI units) $= 1.0$

B (mixed SI units) $= \dfrac{3,600 \times 10^6}{10^6} = 3,600$

B (mixed imperial units) $= 3,600(32.174)^{1/2} = 20,420$

Coefficient of discharge. The coefficient of discharge, K_d, of pressure relief valves expresses the ratio of experimentally to theoretically determined flow through the valve, and must be determined in accordance with code requirements and be certified by statutory authorities. The product of $K_d \times 0.90$ is the coefficient of discharge, K, which may be used in the sizing calculations.

Typical values of K_d for high efficiency pressure relief valves for gas and vapor range from about 0.8 to 0.975, whereby the higher values may have been achieved by oversizing the valve. The engineer designing the pressure relief system must, therefore, consult the manufacturer's data sheets, or catalogs, to determine the value of the coefficient of discharge, K, to be used.

Capacity correction factor K_b. The capacity correction factor, compensates for loss of flow capacity due to back pressure. For conventional safety relief valves operating against constant back pressure,* and balanced safety relief valves operating against constant or variable back pressure, the back-pressure correction factor at 10% overpressure is frequently taken from a curve similar to the one shown in Figure 5-47. Because the value of K_b may vary between valve makes, the valve maker's catalog should be consulted when determining K_b.

The value of K_b becomes more favorable when the valve is sized for an overpressure higher than 10%.

Physical constants. These may be obtained for a variety of gases and vapors from Table B-1 (Appendix B).

The values of the isentropic coefficient k given in this table have been determined at atmospheric pressure and 273K (449R), but may differ at

---

*Conventional safety relief valves are sometimes erroneously claimed to be unaffected by constant back pressure.

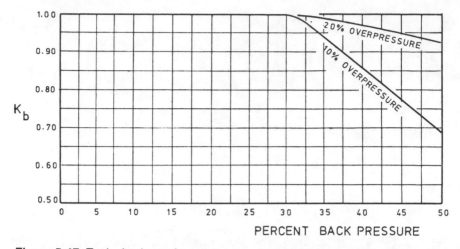

Figure 5-47. Typical values of capacity correction factor K_b due to back pressure in gas or vapor service, applicable to conventional safety relief valves operating against constant back pressure, and balanced safety relief valves operating against constant or variable back pressure.

higher pressures and temperatures. For example, the value of k for air at nominal pressure and temperature is 1.40, but at 10 MPa (1450 lb/in²) and 293K (527R), the value of k is 1.60. If the value of k at inlet conditions is not known, the value at normal pressure and temperature is commonly used.

Compressibility factor Z = PV/RT. This factor compensates for deviations of real gases from the ideal gas laws, and is evaluated at inlet conditions.[36] However, any correction for Z is empirical, as the entire derivation of the equations is based on the ideal gas laws.

If the compressibility factor is not available but the critical pressure and critical temperature are known, approximate values of Z may be obtained from Figures B-1 through B-3 (Appendix B).

If the compressibility factor cannot be determined, a value of 1.0 is commonly used.[36] This value gives conservative results as long as pressure and temperature are not too high.

Sizing for steam. Historically, pressure relief valves for dry saturated steam are sized in the United States by Napier's steam flow equation. The equation has been formulated empirically for conditions of choked flow and reads:

In coherent SI units: $$A = \frac{686\ W}{K\ P_1} \qquad (5\text{-}12a)$$

In mixed SI units:
$$A = \frac{W}{5.25 \ K \ P_1}$$
(5-12b)

In mixed imperial units:
$$A = \frac{W}{51.5 \ K \ P_1}$$
(5-12c)

In the lower pressure range—up to 11 MPa (1,600 lb/in.^{2})—Napier's equation gives similar results to Equation 5-10, using an isentropic exponent of $k = 1.13$. For pressures exceeding 11 MPa (1,600 lb/in.^{2}), however, Napier's sizing equation leads gradually to considerable oversizing.

For pressures between 11 MPa (1,600 lb/in.^{2}) and 22 MPa (3,200 lb/in.^{2}), L. Thompson and O. B. Buxton, Jr., have developed a capacity correction factor (K_n) for use in conjuction with Napier's equation. This factor has been incorporated into the ASME Boiler and Pressure Vessel Code and reads:

For P_1 in terms of Pa abs.:
$$K_n = \frac{27.644 \times 10^{-6} P_1 - 1,000}{33.242 \times 10^{-6} P_1 - 1,061}$$

For P_1 in terms of MPa abs.:
$$K_n = \frac{27.644 \ P_1 - 1,000}{33.242 \ P_1 - 1,061}$$

For P_1 in terms of lb/in.^{2}:
$$K_n = \frac{0.1906 \ P_1 - 1,000}{0.2292 \ P_1 - 1,061}$$

If the steam is superheated, the capacity must, in addition, be corrected by a superheat correction factor (K_{SH}) which is the ratio of maximum isentropic nozzle flow for a given superheat inlet condition to the Napier flow. BS 6759 Part 1: 1984 is one of the standards containing superheat correction factors. One of the national standards following U.S. practice is BS 6759 Part 1: 1984.

The Swiss and German standards, 53, 78 on the other hand, use the basic sizing equation (Equation 5-10) in conjunction with the thermodynamic properties of steam as obtained from the VDI steam table, 55 containing the properties of saturated and superheated steam.

The Swiss standard specifies the following isentropic exponents, as obtained from the VDI steam table:

For saturated steam:

to	2 MPa (	290 lb/in.^{2})	$k = 1.13$
	5 MPa (	720 lb/in.^{2})	$k = 1.10$
	8 MPa (	1,160 lb/in.^{2})	$k = 1.05$
	10 MPa (	1,450 lb/in.^{2})	$k = 1.01$

15 MPa (	2,170 lb/in.^{2})	k = 0.90
20 MPa (	2,900 lb/in.^{2})	k = 0.78
>20 MPa (	>2,900 lb/in.^{2})	k = 0.70

For superheated steam, irrespective of pressure:

to	200°C (	428°F)	k = 1.29
	250°C (	482°F)	k = 1.27
	350°C (	662°F)	k = 1.25
	400°C (	752°F)	k = 1.28
	500°C (	932°F)	k = 1.27
	600°C (	1,112°F)	k = 1.26
	>600°C (	>1,112°F)	k = 1.25

The German standard presents Equation 5-10 in the following form:

$$A = \frac{x\ W}{K\ P_1}$$

in which, for conditions of choked flow:

$$x = \left[\frac{1}{P_1 V_1} k \left(\frac{2}{k+1} \right)^{(k+1)/(k-1)} \right]^{-1/2}$$

A = nominal flow area, mm^2
W = mass rate of flow, kg/h
P_1 = absolute pressure at inlet, bar
V_1 = specific volume at inlet, m^3/kg

The value of x thus calculated may be taken from the graph shown in Figure 5-48.

The x-value of 1.9 represents the corresponding Napier constant. An inspection of Figure 5-48 illustrates the limitations of Napier's equation.

Graphical determination of the nozzle flow area. If the thermodynamic properties of the gas or vapor are known, the nozzle flow area may be found by rigorous thermodynamic calculations using the equation of state for isentropic flow:

$$\frac{v_1^2}{2} + H_1 = \frac{v_n^2}{2} + H_n$$

Neglecting the velocity of approach, then:

$$v_n = [2(H_1 - H_n)]^{1/2}$$

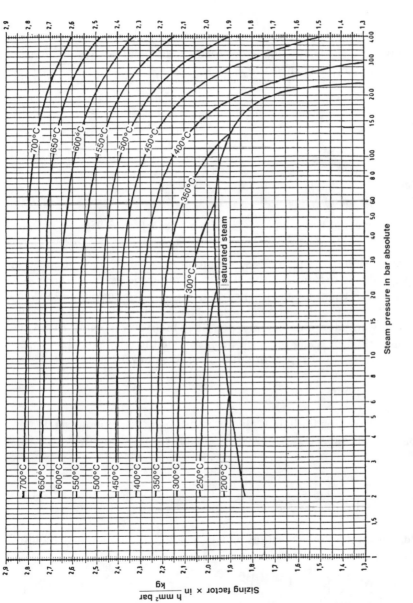

Figure 5-48. Sizing factor x for saturated and superheated steam under conditions of choked nozzle flow. (Taken from the German standard "AD-Merkblatt A2," Edition February 1980, published by Vereinigung der Technischen Ueberwachungsvereine, Rottstr. 17, 43 Essen.)

Thus the theoretical mass flux:

$$G = \frac{1}{V_n} [2(H_1 - H_n)]^{1/2}$$

Substituting this equation into Equation 5-1, the required area at any nozzle section may be calculated from:

$$A_n = \frac{W \, V_n}{B \, K \, [2(H_1 - H_n)]^{1/2}} \tag{5-13}$$

in which

B (coherent SI units) $= 1.0$
B (mixed SI units) $= 3.6$
B (mixed imperial units) $= \dfrac{3,600}{144} (32.174 \times 778.26)^{1/2} = 3,956$

The expansion of a gas in an isentropic process takes place at constant entropy, and this determines the state n. The nozzle sectional area as a function of the distance along the nozzle axis can thus be calculated for any assumed pressure profile along the nozzle axis. Figure 5-49 shows the resulting converging-diverging nozzle profile which corresponds to an assumed linearly decreasing pressure profile along the inlet axis. The nozzle throat area thus found represents the theoretical nozzle area.

Sizing for Liquids

The sizing equation for liquid relief is based on the assumption that no portion of the liquid vaporizes. Under these conditions, the theoretical mass flux is:

$$G = \left(\frac{1}{2 \, (P_1 - P_2) \, \rho} \right)^{-1/2} \tag{5-14}$$

After substituting this equation into Equation 5-1, the nominal required flow area may be calculated from:

$$A_n = \frac{W}{B \, (K \, K_w \, K_v)} \left(\frac{1}{2 \, (P_1 - P_2) \, \rho} \right)^{1/2} \tag{5-15}$$

in which:
   B (coherent SI units) $= 1.0$

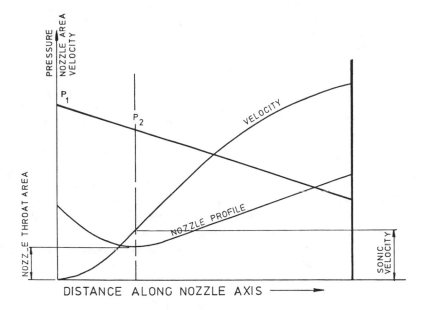

Figure 5-49. Graphical evaluation of nozzle throat area.

$$B \text{ (mixed SI units)} \quad = \frac{3600}{10^6} (10^6)^{1/2} = 3.6$$

$$B \text{ (mixed imperial units)} = \frac{3600}{144} (32.174 \times 144)^{1/2} = 1701.66$$

or

$$A_n = \frac{Q}{B (K K_w K_v)} \left(\frac{\rho}{2 (P_1 - P_2)} \right)^{1/2} \tag{5-16}$$

in which:

$$B \text{ (coherent SI units)} = 1.0$$

$$B \text{ (mixed SI units)} \quad = \frac{60}{10^6} (10^6)^2 = 60 \times 10^{-3}$$

$$B \text{ (mixed imperial units)} = \frac{60}{144} (32.174 \times 144)^{1/2} = 28.36$$

Coefficient of discharge. The coefficient of discharge achieved by safety relief valves in liquid service at an overpressure of 25% is sometimes quoted as $K_d = 0.62$, where $K_d =$ actual coefficient of discharge. However, this value may vary considerably for other types of valves, between valve makes, and between valve sizes of the same make.

As in the case of pressure relief valves in gas or vapor service, the coefficient of discharge used in calculations is derated by 10%, as expressed by the equation

$$K = 0.9 \, K_d \quad \text{or} \quad K = \frac{K_d}{1.1}$$

Capacity correction factor K_w. K_w compensates for loss of flow as a result of diminished valve lift due to back pressure in liquid service. For conventional safety relief valves operating against constant back pressure,* and balanced safety relief valves operating against constant or variable back pressure, the capacity correction factor for conditions of 25% overpressure is frequently taken from a curve similar to the one shown in Figure 5-50.

However, the correction factor may differ between valve makes, and the manufacturer should be consulted for the correct value.

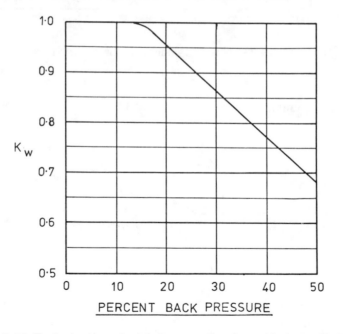

Figure 5-50. Typical values of capacity correction factor K_w due to back pressure in liquid service for conditions of 25% overpressure, applicable to conventional safety relief valves operating against constant back pressure, and balanced safety relief valves operating against constant or variable back pressure.

---

*Conventional safety relief valves are sometimes erroneously claimed to be unaffected by constant back pressure.

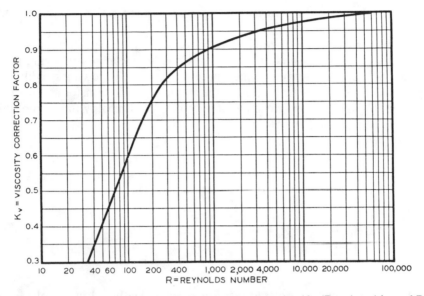

Figure 5-51. Capacity correction factor due to viscosity, K_v. (Reprinted from API RP520 Part I, December 1976, courtesy of American Petroleum Institute.)

The capacity correction factor K_v. If the Reynolds number is less than 60,000, loss of flow through the valve due to viscosity must be compensated for by the capacity correction factor K_v. The values of K_v corresponding to the Reynolds number may be obtained from Figure 5-51. In most practical applications, however, the Reynolds number is higher than 60,000, in which case $K_v = 1.00$.

When a pressure relief valve is sized for viscosity, the valve orifice size should first be determined for nonviscous flow. The next larger standard valve size should then be used to determine the Reynolds number. After this has been done, the factor K_v is obtained from Figure 5-51, and the valve orifice area recalculated for viscous flow. If the valve orifice area thus found is larger than the chosen orifice area, the calculation should be repeated using the next larger standard orifice area.[36]

Sizing for Mixed Phase Flow

Statutory bodies and standard organizations endeavor to put forward firm recommendations on the sizing of pressure relief valves for mixed phase flow. At the time of this writing, these recommendations are still being considered.

In the case of flashing liquids in which the fluid is in the fully liquid state prior to valve opening, pressure relief valves are commonly sized by computing the flow areas separately for vapor and liquid fractions, and then adding these areas together.

The vapor and liquid fractions may be found from the enthalpy function of a throttling process, which may be considered to occur in a relieving pressure relief valve. The process takes place when the fluid flowing steadily in a pipeline passes a flow restriction in the absence of heat transfer and without appreciable kinetic energy change of the fluid. The enthalpy function of this process is expressed by:

$$H_1 = H_2$$

or, for a liquid that flashes from a fully liquid state:

$$\begin{aligned}
H_{\ell,1} &= H_{(\ell+v),2} \\
&= x\, H_{v,2} + (1-x)H_{\ell,2} \\
&= x(H_{v,2} - H_{\ell,2}) + H_{\ell,2}
\end{aligned}$$

where

$H_{\ell,1}$	= enthalpy of saturated or supercooled liquid in vessel
$H_{(\ell+v),2}$	= enthalpy of liquid/vapor mixture downstream of nozzle
$H_{\ell,2}$	= enthalpy of saturated liquid downstream of nozzle
$H_{v,2}$	= enthalpy of saturated vapor downstream of nozzle
$H_{v,2} - H_{\ell,2}$	= enthalpy of vaporization
x	= fraction

Therefore:

$$x = \frac{H_{\ell,1} - H_{\ell,2}}{H_{v,2} - H_{\ell,2}} \qquad (5\text{-}17)$$

$$W_v = xW \qquad (5\text{-}18)$$

$$W_\ell = (1-x)W \qquad (5\text{-}19)$$

Deviating from the above presentation, the enthalpy state 2 is more frequently based on the critical pressure:

$$P_n = P_1\left(\frac{2}{k+1}\right)^{k/(k+1)}$$

Examples of the different approaches are the Swiss Code of Practice 78, which requires the enthalpy state 2 to be based on the atmospheric pressure; while the German Code of Practice 79 requires the enthalpy state 2 to be based on the critical pressure. The latter code, however, requires also that the sum of the flow areas thus found be multiplied by a factor of 1.2 to cover deviations of the actual flow from the theoretical flow when using the proposed calculating method.

The ASME Code, issue 1983, Division 1, Appendix 11, proposes a sizing method which applies to saturated water only. This sizing method consists of a graph, from which the flow capacity per unit area can be read for a given upstream pressure. The graph, however, applies only to pressure relief valves with a converging nozzle type throat, which exhibits a coefficient of discharge, K_d, in excess of 0.90.

Where pipelines and vessels contain a vapor-liquid mixture, pressure relief valves are commonly sized for conditions of dry saturated vapor.

Calculation of Pressure Loss in Inlet Pipeline in Gas or Vapor Service

In the case of direct-acting pressure relief valves, the pressure loss in the pipeline between the protected equipment and the inlet connection of the valve should not exceed 3% of the set pressure, as described on page 181.

The following equation for determining the inlet pipeline pressure loss is based on conditions of choked nozzle flow:

$$\Delta P = \left(\Sigma \ \varsigma + f \ \frac{1}{d} \right) \left(\frac{K_d \ K_b \ A_n}{A_i} \right)^2 \frac{P_1}{2Z} \ k \left(\frac{2}{k+1} \right)^{(k+1)/(k-1)} \tag{5-20}$$

The equation applies in any consistent set of units. Its derivation can be found in Appendix A.

Calculation of Back Pressure in Gas or Vapor Service

The back pressure P_2 at the outlet flange of pressure relief valves consists of the built-up back pressure due to friction of the discharging fluid in the discharge pipeline, and the pressure P_t at the outlet of the discharge pipeline.

Accordingly, the calculation proceeds by first calculating the pressure at the terminus of the pipeline for known conditions at the inlet of the pressure relief valve, and then adding the pipeline friction loss, to arrive at the back pressure at the outlet of the pressure relief valve.

The following equations are based on conditions of choked flow through the pressure relief valve and on the assumption that flow through the pipeline is isothermal. They apply in any consistent set of units. Derivations of Equations 5-21 and 5-22 can be found in Appendix A.

Terminal pressure P_t.

$$P_t = \frac{K_d K_b A_n P_1}{A_t} \left(\frac{2}{k+1} \right)^{k/(k-1)} \left(\frac{1}{Z} \right)^{1/2} \tag{5-21}$$

If the calculated value of $P_t \geqslant P_3$, the discharge velocity is sonic. If the calculated value of $P_t < P_3$, the discharge velocity is subsonic, and the actual value of $P_t = P_3$.

Back pressure P_2.

$$P_2 = \left[P_t^2 + \left(2 \ln \frac{P_2}{P_t} + \Sigma K_f + \frac{fl}{d_t} \right) \right.$$

$$\left. \left(\frac{K_d K_b A_n P_1}{A_t} \right)^2 \frac{k}{Z} \left(\frac{2}{k+1} \right)^{2k/(k-1)} \right]^{1/2} \tag{5-22}$$

Sizing of Diffuser-Type Silencers for Pressure Relief Valves

The design principles of diffuser-type silencers are discussed in Chapter 2, page 40. Figure 5-46 shows the schematic arrangement of this type of silencer mounted in the discharge pipeline of a direct-acting pressure relief valve. The calculation proceeds by the following steps:

1. Calculate diffuser stack area, A_t.
2. Calculate back pressure at outlet of diffuser nozzles, P_s.
3. Calculate diffuser nozzle area, A_{sn}.
4. Calculate pressure at inlet of diffuser nozzles, P_4.
5. Calculate back pressure at section 6 (Figure 5-46) of discharge pipe, P_5.
6. Check pressure at section 6 (Figure 5-46) of discharge pipe, P_6.
7. Calculate back pressure at outlet of pressure relief valve, P_2.

The following equations are based on conditions of choked flow through the pressure relief valve and on the assumption that nozzle flow is isentropic pipeline flow isothermal. They apply in any consistent set of units. Derivations of Equations 5-23 through 5-26 are given in Appendix A.

Diffuser stack area A_t. For Mach number $N_t < 1.0$:

$$A_t = \frac{K_d K_b A_n P_1}{P_3 N_t} \left(\frac{2}{k+1} \right)^{k/(k-1)} \left[\frac{1}{Z \left(1 + \frac{k-1}{2} N_{sn}^2 \right)} \right]^{1/2} \tag{5-23}$$

The Mach number at the stack outlet may be chosen as $N_t \leqslant 0.3$, and at the outlet of the diffuser nozzles as $N_{sn} \leqslant 0.9$.

Back pressure P₅. At section 6 (Figure 5-48) of discharge pipe:

$$P_s = P_3 + \left(\Sigma\varsigma + \frac{fl}{d_t}\right)\left(\frac{K_d K_b A_n P_1}{A_t}\right)^2$$

$$\frac{k}{2P_3 Z\left(1 + \dfrac{k-1}{2}N_{sn}^2\right)}\left(\frac{2}{k+1}\right)^{2k/(k-1)} \tag{5-24}$$

where
   $1 = $ length of diffuser stack

Diffuser nozzle area Aₛₙ. For Mach number $N_{sn} < 1.0$:

$$A_{sn} = \frac{K_d K_b A_n P_1}{N_{sn} P_s K_z}\left(\frac{2}{k+1}\right)^{k/(k-1)}\left[\frac{1}{Z\left(1 + \dfrac{k-1}{2}N_{sn}^2\right)}\right]^{1/2} \tag{5-25}$$

where
   $K_z = $ coefficient of discharge of diffuser nozzle

Its value may be taken as 0.6 for nozzles with sharp entry, and as 0.8 for nozzles with rounded entry.
   The diameter of the diffuser nozzles may be taken as 5mm or 3/16 in., and the open area ratio of the diffuser plate taken as between 0.1 and 0.3. If the nozzles cannot be accommodated in a flat plate, a cone- or bucket-shaped diffuser may be used.

Pressure P₄. At inlet of diffuser nozzles:

$$P_4 = \frac{P_s}{\left(1 - \dfrac{k-1}{k+1}N_{sn}^2\right)^{k/(k-1)}} \tag{5-26}$$

Back pressure P₅. At section 6 (Figure 5-48) of discharge pipe:

$$P_5 = \left[P_4^2 + \left(2\ln\frac{P_5}{P_4} + \varsigma + \frac{fl}{d_4}\right)\right.$$

$$\left.\left(\frac{K_d K_b A_n P_1}{A_4}\right)^2\frac{k}{Z}\left(\frac{2}{k+1}\right)^{2k/(k-1)}\right]^{1/2} \tag{5-27}$$

where
  ℓ = straight length of pipe between sections 4 and 6
  ζ = resistance coefficient of pipe enlargement

The resistance coefficient of a sudden enlargement in terms of velocity heads at the outlet may be taken as:

$$\zeta = \left[1 - \left(\frac{d_6}{d_4}\right)^2\right]^2 \left(\frac{d_4}{d_6}\right)^2$$

Note: Equation 5-27 is identical to Equation 5-22 except for subscripts.

Check pressure P$_6$. At section 6 (Figure 5-46) of discharge pipe:

$$P_6 = \frac{K_d K_b A_n P_l}{A_6}\left(\frac{2}{k+1}\right)^{k/(k-1)}\left(\frac{1}{Z}\right)^{1/2} \qquad (5\text{-}28)$$

If the calculated value of $P_6 \geqslant P_5$, the velocity at section 6 is sonic. If the calculated value of $P_6 < P_5$, the velocity at section 6 is subsonic, and the actual value of $P_6 = P_5$.
Note: Equation 5-28 is identical to Equation 5-21 except for subscripts.

Back pressure P$_2$. At the outlet of the pressure relief valve:

$$P_2 = \left[P_6^2 + \left(2\ln\frac{P_2}{P_6}\Sigma\zeta + \frac{f\ell}{d_6}\right)\left(\frac{K_d K_b A_n P_l}{A_6}\right)^2\right.$$

$$\left.\frac{k}{Z}\left(\frac{2}{k+1}\right)^{2k/(k-1)}\right]^{1/2} \qquad (5\text{-}29)$$

where
  ℓ = length of straight pipe section between sections 2 and 6
  ζ = resistance coefficient of pipe bend

Note: Equation 5-29 is identical to Equation 5-22 except for subscripts.
If the back pressure thus calculated exceeds the permissible back pressure for the valve, the pipe size must be increased and/or the pressure drop across the diffuser lowered.

Discharge Reactive Force

The general equation for determining the discharge reactive force is, in terms of SI units:

$$F = Wv_t + A_t (P_t - P_3)$$

in which, for gases or vapors, the value of P_t may be found from Equation 5-21. This determines whether the discharge velocity is subsonic or sonic. The calculated discharge reactive force should be multiplied by a dynamic load factor, as described in Reference 56, to allow for the shock loading on valve opening.

Equations 5-30 and 5-31 have been developed for conditions of choked flow through the pressure relief valve and on the assumption of isothermal pipeline flow. They apply in any consistent set of units. Derivations of the equations can be found in Appendix A.

Discharge reactive force for gas or vapor for condition of subsonic discharge velocity.

$$F = \frac{(K_d K_b A_n P_1)^2}{A_t P_3 Z} k \left(\frac{2}{k+1}\right)^{2k/(k-1)} \tag{5-30}$$

Discharge reactive force for gas or vapor for condition of sonic discharge velocity.

$$F = (1 + k) \left[K_d K_b A_n P_1 \left(\frac{2}{k+1}\right)^{k/(k-1)} \left(\frac{1}{Z}\right)^{1/2} \right] - A_t P_3 \tag{5-31}$$

NUMERICAL EXAMPLES

Fluid data:
  Nitrogen gas
  $M = 28.106$
  $k = 1.4$
  $P_c = 3.38 \times 10^6 Pa$
  $T_c = 126.3K$

Flow data:
  $W = 5.85$ kg/s
  set pressure $= 30 \times 10^6 Pa$ (gauge)
  overpressure $= 10\%$
  $P_1 = 33.1013 \times 10^6 Pa$ (absolute)
  $T_1 = 200°C = 473K$

Valve type:
Conventional safety relief valve
$K_d = 0.975$
$K = 0.9 K_d = 0.878$

Discharge to atmosphere. Therefore:
$P_3 = 101.3 \times 10^3 Pa$

Case 1: Discharge pipeline of uniform cross section.
Case 2: Discharge pipeline with diffuser-type silencer.

Calculate valve size. Use Equation 5-11.

Reduced pressure

$$P_1/P_c = \frac{33.1013}{3.38} = 9.79$$

Reduced temperature

$$T_1/T_c = \frac{473}{126.3} = 3.76$$

From Figure B-2:
$Z = 1.19$

Then:

$$A_n = \frac{5.85}{0.878 \times 33.1013 \times 10^6}$$

$$\left[\frac{28.016}{10^3 \times 8.3143 \times 473 \times 1.19} \times 1.4 \left(\frac{2}{2.4}\right)^{2.4/0.4}\right]^{-1/2}$$

$$= 120.1 \times 10^{-6} \ m^2$$

The appropriate valve size to API Std 526 (Nov. 1969) is 1½ E 2½, for which $A_n = 126.5 \times 10^{-6} m^2$, inlet flange DN40 (NPS 1½), and outlet flange DN65 (NPS 2½).

Calculate pressure loss in inlet pipeline. Let:

$d_i = 40.9 \times 10^{-3} m$
$A_i = 1.313 \times 10^{-3} m^2$

Length of straight pipe is 5.0m, and allow one 90° bend with $r = 3d$ ($\zeta \times 0.25$), a sharp pipe entry ($\zeta = 0.5$), and a pipe friction factor of $f = 0.02$.

Then from Equation 5-20:

$$\Delta P = \left(0.75 + \frac{0.02 \times 5.0 \times 10^3}{40.9}\right) \left(\frac{0.975 \times 126.5}{1.313 \times 10^3}\right)^2$$

$$\frac{33.1013 \times 10^6}{2 \times 1.19} \times 1.4 \left(\frac{2}{2.4}\right)^{2.4/0.4}$$

$$= 183.8 \times 10^3 Pa$$

Permissible: 3% of $33.1013 \times 10^6 = 993 \times 10^3 Pa$

Calculate back pressure P_2 for case 1. Given a discharge pipeline of uniform cross section, let:

$$d_2 = d_t = 62.7 \times 10^{-3} m$$
$$A_2 = A_t = 3.088 \times 10^{-3} m^2$$

Length of straight pipeline $\ell = 18m$, and allow one 90° bend with $r = 3d$ ($\zeta = 0.25$), and a pipe friction factor $f = 0.02$. From Equation 5-21:

$$P_t = \frac{0.975 \times 126.5 \times 33.1013 \times 10^3}{3.088} \left(\frac{2}{2.4}\right)^{1.4/0.4} \left(\frac{1}{1.19}\right)^{1/2}$$

$$= 640.3 \times 10^3 Pa$$

The calculated pressure P_t is higher than atmospheric, so the outlet velocity is sonic. Then from Equation 5-22:

$$P_2 = \left[640.3^2 \times 10^6 + \left(2 \ln \frac{P_2}{640.3 \times 10^3} + 0.25 + \frac{0.02 \times 18 \times 10^3}{62.7}\right)\right.$$

$$\left.\left(\frac{0.975 \times 126.5 \times 33.1013 \times 10^3}{3.088}\right)^2 \frac{1.4}{1.19} \left(\frac{2}{2.4}\right)^{2.8/0.4}\right]^{1/2}$$

$$= 2.305 \times 10^6 Pa$$

Permissible back pressure is 15% of a set pressure. Thus:

$$P_2 \text{ (permissible)} = 0.15 \times 33.1013 \times 10^6 = 4.965 \times 10^6 Pa.$$

Calculate back pressure for case 2. Given a diffuser-type silencer, the calculation is carried out in steps as described on page 216.

a. **Stack area A_t.** Mach number for flow from stack:
$N_t = 0.3$

Hence:
$P_t = P_3 = 101.3 \times 10^3 \text{Pa}$

Mach number for flow from diffuser nozzles:
$N_{sn} = 0.9$

Then from Equation 5-23:

$$A_t = \frac{0.975 \times 126.5 \times 33.1013}{10^6 \times 0.3 \times 0.1013} \left(\frac{2}{2.4}\right)^{1.4/0.4} \left[\frac{1}{1.19\left(1 + \frac{0.4}{2} 0.9^2\right)}\right]^{1/2}$$

$$= 60.4 \times 10^{-3} \text{m}^2$$

Hence:
$d_t \, (\text{min}) = 277.2 \times 10^{-3} \text{m}$

Chose pipe DN300 (NPS 12), with:
$d_t = 305 \times 10^{-3} \text{m}$
$A_t = 72.97 \times 10^{-3} \text{m}^2$

b. **Back pressure P_s at outlet of diffuser nozzles.** Let the length of diffuser stack be 6m, the pipe friction factor $f = 0.02$, and the Mach number for flow from diffuser nozzles $N_{sn} = 0.9$. Then from Equation 5-24:

$$P_s = 101.3 \times 10^3 + \frac{0.02 \times 6 \times 10^3}{305} \left(\frac{0.975 \times 126.5 \times 33.1013 \times 10^3}{72.97}\right)^2$$

$$\frac{1.4}{2 \times 101.3 \times 10^3 \times 1.19 \left(1 + \frac{0.4}{2} 0.9^2\right)} \left(\frac{2}{2.4}\right)^{2.8/0.4}$$

$$= 103.0 \times 10^3 \text{Pa}$$

c. **Diffuser nozzle area A_{sn}.** Let $K_z = 0.6$ for diffuser nozzle with sharp edged entry. Then from Equation 5-25:

$$A_{sn} = \frac{0.975 \times 126.5 \times 33.1013}{0.9 \times 103.0 \times 10^3 \times 0.6} \left(\frac{2}{2.4}\right)^{1.4/0.4} \left[\frac{1}{1.19\left(1 + \frac{0.4}{2}\, 0.9^2\right)}\right]^{1/2}$$

$$= 32.98 \times 10^{-3} \text{m}^2$$

Open area ratio for flat-plate diffuser:

$$\sigma = \frac{A_{sn}}{A_t} = \frac{32.98}{72.97} = 0.45$$

The maximum open-area ratio should not exceed 0.3. This open-area ratio may be achieved by making the diffuser cone or bucket shaped.

d. **Pressure P_4 at inlet of diffuser nozzles.** From Equation 5-26:

$$P_4 = \frac{103.0 \times 10^3}{\left(1 - \frac{0.4}{2.8}\, 0.9^2\right)^{1.4/0.4}} = 158.4 \times 10^3 \text{Pa}$$

e. **Back pressure P_5 at section 6 (Figure 5-46) of discharge pipe.** Calculate the resistance coefficient of pipe enlargement as described in Chapter 5, page 218.

Small diameter of enlargement $d_6 = 62.7 \times 10^{-3}$m
Large diameter of enlargement $d_4 = 305 \times 10^{-3}$m

Then:

$$\zeta = \left[1 - \left(\frac{62.7}{305}\right)^2\right]^2 \left(\frac{305}{62.7}\right)^2 = 21.7$$

Let
straight length of pipe $\ell = 0.5$m
pipe friction factor $f = 0.02$
$A_4 = A_t$

Then from Equation 5-27:

$$P_5 = \left[158.4^2 \times 10^6 + \left(2 \ln \frac{P_5}{158.4 \times 10^3} + 21.7 + \frac{0.02 \times 0.5 \times 10^3}{305}\right)\right.$$

$$\left. \left(\frac{0.975 \times 126.5 \times 33.1013 \times 10^3}{72.97}\right)^2 \frac{1.4}{1.19} \left(\frac{2}{2.4}\right)^{2.8/0.4}\right]^{1/2}$$

$$= 219.3 \times 10^3 \text{Pa}$$

f. Check pressure P_6 at section 6 (Figure 5-46) of discharge pipe. Let $A_6 = 3.088 \times 10^{-3} \text{m}^2$. Then from Equation 5-28:

$$P_6 = \frac{0.975 \times 126.5 \times 33.1013 \times 10^3}{3.088} \left(\frac{2}{2.4}\right)^{1.4/0.4} \left(\frac{1}{1.19}\right)^{1/2}$$

$$= 640.3 \times 10^3 \text{Pa}$$

Since $P_6 > P_5$, velocity at outlet of the pipe to the diffuser is sonic.

g. Back pressure P_2 at outlet of pressure relief valve. Let

$$d_6 = 62.7 \times 10^{-3} \text{m}$$
$$A_6 = A_2 = 3.088 \times 10^{-3} \text{m}^2$$

Also given straight length of pipe $\ell = 6$m, one bend with $r = 3d$, and $\zeta = 0.25$, $f = 0.02$. Then from Equation 5-29:

$$P_2 = \left[640.3^2 \times 10^6 + \left(2 \ln \frac{P_2}{640.3 \times 10^3} + 0.25 + \frac{0.02 \times 6 \times 10^3}{62.7}\right)\right.$$

$$\left. \left(\frac{0.975 \times 126.5 \times 33.1013 \times 10^3}{3.088}\right)^2 \frac{1.4}{1.19} \left(\frac{2}{2.4}\right)^{2.8/0.4}\right]^{1/2}$$

$$= 1.656 \times 10^6 \text{Pa}$$

The maximum permissible back pressure for conventional safety relief valves may be taken as 15% of the set pressure.

$$P_2 \text{ (permissible)} = 33.1013 \times 10^6 \times 0.15$$
$$= 4.965 \times 10^6 \text{Pa}$$

Thus, the back pressure is well within its permissible limit.

Calculate discharge reactive force for case 1. Discharge pipeline:

$$d_2 = d_t = 62.7 \times 10^{-3}m$$

and

$$A_2 = A_t = 3.088 \times 10^{-3}m^2$$

Determine terminal pressure P_t from Equation 5-21:

$$P_t = \frac{0.975 \times 126.5 \times 33.1013 \times 10^3}{3.088} \left(\frac{2}{2.4}\right)^{1.4/0.4} \left(\frac{1}{1.19}\right)^{1/2}$$

$$= 640.3 \times 10^3 Pa$$

Because the terminal pressure is higher than the atmospheric pressure, the flow velocity at the discharge pipe outlet is sonic. Then from Equation 5-31:

$$F = (1 + 1.4) \left[0.975 \times 126.5 \times 33.1013 \left(\frac{2}{2.4}\right)^{1.4/0.4} \left(\frac{1}{1.19}\right)^{1/2}\right]$$

$$- 3.088 \times 101.3$$

$$= 4432N$$

Calculate discharge reactive force for case 2. Diffuser stack is designed for subsonic flow velocity. Size of diffuser stack:

$$d_t = 305 \times 10^{-3}m$$

and

$$A_t = 72.97 \times 10^{-3}m^2$$

Then from Equation 5-30:

$$F = \frac{(0.975 \times 126.5 \times 33.1013)^2}{72.97 \times 101.3 \times 1.19} 1.4 \left(\frac{2}{2.4}\right)^{2.8/0.4}$$

$$= 740.4N$$

The above calculated discharge reactive forces apply to steady state flow only. To account for the effects of the suddenly applied reactive force upon valve opening, a dynamic load factor as recommended in Reference 56 may be applied.

6

Rupture Discs

Rupture discs are the pressure sensitive element of non-reclosing pressure relief devices for the protection of fluid systems against damage from excessive overpressure or reverse pressure. When the pressure differential across the rupture disc reaches a predetermined level, the disc will burst, relieving the system from excessive overpressure or reverse pressure. If the fluid is a gas and the disc has burst as a result of excessive overpressure, the system may lose nearly all its contents. To make the fluid system operable again, the rupture disc must be replaced.

Most rupture discs require a separate holder, consisting of inlet and outlet holder parts between which the disc is clamped. Rupture disc and holder together represent the rupture disc device.

The construction material of rupture discs covers a range of ductile and brittle materials. Based on these construction materials, a wide range of rupture disc constructions have evolved to satisfy a multitude of operational requirements.

The original rupture disc consisted of a plain flat ductile metal disc that was clamped between two flanges. When the disc was subjected to rising pressure, it would stretch and form into a hemispherical dome prior to bursting.

This type of rupture disc was soon abandoned for a prebulged version, in which the dome was preformed by applying a pressure to the disc somewhat higher than the operating pressure.

However, the prebulged disc still is limited in its range of application. To overcome these limitations, a variety of other types of rupture discs has been developed, some of which fail in compression instead of in tension.

The following types of ductile metal rupture discs are in common use:

- Prebulged, solid construction
- Prebulged, composite construction
- Flat, composite construction
- Reverse buckling

Low pressure rupture discs for the release of large amounts of gases resulting from the deflagration of dust, flammable gases and vapors as may occur in flower mills, grain silos, ducts, dryers, etc. are referred to as vent panels. Their design is generally based on flat and sometimes on prebulged composite rupture discs.

The types of rupture discs in use made of brittle materials are:

- Flat diaphragm
- Reverse buckling

Terminology

Table 6-1 lists some important terms used in the United States and gives their counterparts in British usage.

Table 6-1
Rupture Disc Terminology

USA	BS 2915:1984
Rupture disc	Bursting disc
Back pressure	Reverse pressure
Vacuum condition	Reverse pressure
Prebulged rupture disc	Predomed bursting disc
Composite rupture disc	Slotted lined bursting disc
Reverse buckling disc	Reverse domed bursting disc
Heat shield	Temperature shield
Vacuum support	Reverse pressure support
Back pressure support	Reverse pressure support
Lot	Batch

Definitions

Rupture disc device: A non-reclosing pressure relief device, consisting of rupture disc and holder, that is actuated by differential pressure and functions by bursting of the rupture disc.

Rupture disc: The pressure containing and pressure sensitive element of the rupture disc device.

Prebulged rupture disc: A rupture disc that is prebulged in the direction of the fluid pressure.

Prebulged composite rupture disc: A prebulged rupture disc consisting of two or more layers of which one is slotted so as to reduce its strength and to control the bursting pressure.

Reverse buckling disc: A rupture disc that is domed against the direction of the fluid pressure.

Holder: That component of a rupture disc device which holds the rupture disc around its circumference, consisting of the inlet and outlet holder parts.

Vent panel: A low pressure venting device which consists of large area vent panels. It is used in dust, gas, or vapor handling systems to vent the almost instantaneous volumetric and pressure changes resulting from dust, gas, or vapor deflagration.

Vacuum support: A device that supports the rupture disc against collapse due to vacuum conditions.

Back pressure support: A device that supports the rupture disc against collapse due to superimposed back pressure.

Heat shield: A device that protects the rupture disc from a heat source in a manner which does not interfere with rupture disc operation.

Coincident temperature: The temperature used in conjunction with a bursting pressure.

Bursting pressure: The value of the pressure differential across the rupture disc at which a rupture disc device functions.

Stamped (rated) bursting pressure: The bursting pressure which is shown (stamped) on the tag of the rupture disc. This pressure is coincident to the temperature shown on the tag.

Burst pressure tolerance: The maximum variation in bursting pressure that the disc may have from its stamped bursting pressure rating, expressed as a percentage of the stamped rating. The burst pressure tolerance is generally stated as a plus/minus percentage of the stamped (rated) pressure.

Allowable operating pressure: This is the maximum operating pressure at which the rupture disc should be allowed to operate so as to achieve an acceptable service life.

Operating pressure margin: This is the difference between the bursting pressure and the operating pressure of the rupture disc.

Manufacturing range: An allowable range of pressures applied by the manufacturer around the specified bursting pressure within which a rupture disc can be rated or stamped.

Lot: A quantity of rupture discs made as a single group of the same type, size, and limits of bursting pressure and coincident temperature, manufactured from material of the same identity and properties.

Deflagration: Burning that takes place at a flame speed below the velocity of sound in the unburned medium.

Detonation: Burning that takes place at a flame speed above the velocity of sound in the unburned medium.

Explosion: A bursting of a building or container as a result of development of internal pressure beyond the confinement capacity of the building or container.

Applications of Rupture Discs

Because rupture discs do not reclose after bursting, the decision to install rupture discs for the relief of overpressure can have important economical consequences.

However, there are many applications where rupture discs are likely to perform better than pressure relief valves, as in the following:

- Under conditions of uncontrolled reaction or rapid overpressurization where the inertia of a pressure relief valve would inhibit the required rapid release of excess pressure.
- Where even minute leakage of the fluid to the atmosphere cannot be tolerated at normal operating conditions.
- Where the fluid is extremely viscous.
- Where the fluid would tend to deposit solids to the underside of the pressure relief valve that would render the valve inoperable.
- Where low temperature would cause pressure relief valves to seize.

Rupture discs may fill special requirements when mounted in parallel or series.

Mounted in parallel:

- Where one disc alone cannot satisfy the relief requirements.
- Where one disc serves as the primary pressure relief device while the second disc, set at a higher pressure, protects the pressure system against unusual overpressure conditions.
- In conjunction with a switch-over mechanism which puts one section of the pressure relief system on-line while isolating the second pressure re-

lief system. When the first section has failed or needs to be serviced, the second section may be put into service by operating the switch-over valve. The latter must be so designed that the second port is open before the first port closes.

Two discs in series:

- To prevent fluid discharge to the atmosphere due to premature failure of the inner disc as a result of fatigue or corrosion.
- To serve as a remotely operated quick-opening device. The discs are designed in this case for a fraction of the system pressure but are prevented from bursting by pressurizing the space between the discs. Upon dumping the pressure between the discs, the discs will burst within a few milliseconds.

Rupture discs may also be mounted in conjunction with pressure relief valves in the following manner:

1. In parallel with a pressure relief valve where the pressure relief valve may serve as the primary pressure relief device while the rupture disc, set at a higher pressure, protects the pressure system against unusual overpressure conditions.
2. In series upstream of a pressure relief valve for the following purposes:
    - To prevent leakage past the closed valve disc to the atmosphere or vent system.
    - To prevent deposits from forming around the valve disc that would impair the operation of the pressure relief valve.
    - To prevent corrosive fluids from leaking into the pressure relief valve. This may allow the pressure relief valve to be made of standard construction materials.
    - To reduce the cost of maintaining pressure relief valves.
3. In series downstream of the valve for the following purposes:
    - To prevent leakage to the vent system under normal operating conditions.
    - To prevent corrosive fluids in the vent system from entering the valve.
4. In series upstream and downstream of the pressure relief valve to combine the advantages of upstream and downstream installation of rupture discs.

Rupture Discs in Gas and Liquid Service

The ASME Code, Section VIII, Division 1, UG-127, contains the following footnote, applying to ductile metal rupture discs:

Application of rupture disc devices to liquid service should be carefully evaluated to assure that the design of the rupture disc device and the dynamic energy of the system on which it is installed will result in sufficient opening of the rupture disc.

When ductile metal rupture discs burst in gas service, the expanding gas forces the disc open in milliseconds. The actual opening area, however, may vary between types of discs of identical size made of the same construction material.

When used in liquid service, ductile metal rupture discs will burst in the same manner only if there is a large enough gas pocket between the liquid and the rupture disc. Advice on the volume of gas that must be maintained for this purpose must be obtained from the manufacturer. Typically, the minimum gas volume to be maintained should be equivalent to at least 10 diameters of pipe to which the rupture disc is connected.

If the system is totally full of liquid and the pressure increase is due to thermal expansion of the liquid or process reaction, the pressure rise decays initially in response to an increase in the system volume due to deformation of the rupture disc. With ductile metal rupture discs of the prebulged tension type, this will occur with growth of the dome. With ductile metal rupture discs of the buckling type, this will occur with initial collapse of the dome and may leave a partially reversed disc.

If the liquid volume continues to rise, either as a result of continued thermal expansion of the liquid, or process reaction or pump action, only tension-type prebulged discs and some specially designed reverse acting discs, as discussed under the reverse buckling disc section commencing on page 236, may be used in totally full liquid systems. Cross-scored reverse buckling discs and all reverse buckling discs with knife blades, which rely for bursting on snap reverse action and have been partially reversed, may burst at a pressure considerably higher than the set pressure, resulting possibly in the rupture of the pressure system. For these reasons, these discs should be used in liquid systems only if the presence of a critical air pocket below the disc can be assured.

The selection of ductile metal rupture discs for totally full liquid systems is therefore complex. For this reason, the user should consult the manufacturer on selection and installation of rupture discs in totally full liquid systems.

Graphite rupture discs, on the other hand, give instantaneously full opening upon bursting, irrespective of the type of service.

Independent of the type of rupture disc used in liquid service, however, a gas pocket should be maintained for other reasons. The gas pocket minimizes pressure rises due to volume change of the liquid and dampens peak impulse loads such as from waterhammer, resulting in a reduction in the frequency of disc failure.

Pressure vessels on road or rail containing liquids present still another problem. If sloshing liquid can hit the disc, the hydraulic load imposed on the disc might be high enough to burst the disc. To combat sloshing of the liquid, most tankers are provided with baffles.

Locating Rupture Discs

Rupture discs should be placed as close to the vessel or source of pressure as possible to ensure the fastest response possible of the rupture disc to overpressure. This consideration is particularly important in installations where the overpressure build-up is extremely fast.

Construction Materials of Rupture Discs

The most common ductile materials for rupture discs are:

Stainless steel
Aluminum
Inconel
Nickel
Monel
Hastelloy B and C

Other ductile materials include:

Tantalum
Platinum
Gold
Silver
Titanium

Rupture discs of ductile material may also be provided with protective coatings or linings. Common coatings are:

FEP, TFE or PFE
Epoxy
Vinyl
and common linings:
FEP, TFE or PFA
Lead
Polyethylene

Table 6-2 gives maximum recommended temperatures for various metals, coatings, and linings.

Table 6-2a
Recommended Maximum Temperatures

Aluminum	125°C	260°F
Silver	125°C	260°F
Nickel	425°C	800°F
Monel	425°C	800°F
Inconel	535°C	1000°F
316 Stainless Steel	480°C	900°F

Table 6-2b
Maximum Temperatures for Coatings and Linings

Lead	120°C	250°F
Polyvinylchloride	80°C	180°F
FEP	215°C	400°F
TFE or PFA	260°C	500°F

Taken from "Rupture Disc Technical Manual," March 1, 1985, Continental Disc Corporation.

Rupture discs of brittle material are made almost entirely of graphite, although cast iron and porcelain have been used or tried.

The graphite commonly used for rupture discs is made from low ash petroleum cokes, calcinated at high temperatures. It is then mixed with pitch, formed into blocks, and heat-treated. The result is a porous, brittle material that requires sealing for use in rupture discs. This is commonly done by impregnating the graphite under vacuum with either phenolic or furane resins.

Less frequently, pure graphite is used. This is exfoliated graphite, originally in powder form. When suitably compressed, the graphite forms into an impervious sheet. However, pure graphite will in some instances absorb some liquid. This problem can be overcome in these cases by applying a suitable coating to the disc on the process side only.

Temperature and Bursting Pressure Relationships

Temperature influences the strength of materials so that there is a relationship between temperature and bursting pressure. This relationship may vary between rupture discs of identical material but different construction. Figure 6-1 shows such a relationship for solid metal prebulged rupture discs, as provided by one manufacturer. Specific relationships are derived by the manufacturer with each lot of material.

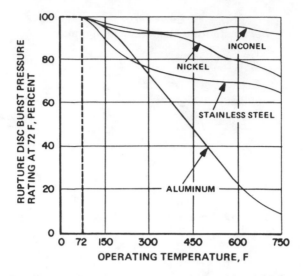

Figure 6-1. Temperature/bursting pressure relationship of CDC solid prebulged rupture discs made of a variety of materials. (Courtesy of Continental Disc Corporation.)

In the case of resin-impregnated graphite discs, the resin limits the operating temperature to about 180°C (350°F). At 150°C (300°F), the bursting pressure would be lower by 10 to 15% than at 20°C (68°F).

Efforts to raise the permissible operating temperature of these discs led to the development of a construction in which the main body is left unimpregnated while an impervious pure graphite foil provides the seal. Such discs are capable of satisfactory service to 350°C (600°F).

Pure graphite discs are particularly suitable in high temperature environments, depending on the degree of oxidation by the process fluid. In fully oxidizing environments, 350°C (660°F) is regarded as a maximum, whereas in fully inert systems 3000°C (5400°F) is considered possible.

The effect of temperature on the strength of reverse buckling discs made of pure graphite has been established by prolonged temperature testing. Between 150°C (300°F) and 600°C (1110°F), the temperature/strength relationship is almost directly proportional. At 150°C (300°F), the strength is 90% that of a disc at 20°C (68°F) and, at 600°C (1110°F), 75%.

The actual rupture disc temperature is often difficult to establish. Because rupture discs are normally located in a pocket or pipe where there is no flow, the temperature of the rupture disc is considerably lower than that of the fluid. Other factors influencing the disc temperature are heat loss through piping and flanges, and the distance of the rupture disc from the actual heat source.

Manufacturers request the user to estimate the disc temperature. If necessary, the installation must be corrected after the system has been in use.

In some cases, however, it may be permissible to ignore the effect of higher temperature on the bursting pressure. This procedure turns an uncertainty into a safety margin. But if the temperature can drop substantially below the ambient, the manufacturer may have to test the rupture disc at that low temperature to determine the bursting pressure.

Heat Shields

Heat shields are devices which are mounted ahead of the rupture disc to shield the disc from heat radiation, or heat radiation and convection. Such devices must be designed and mounted in a manner which does not interfere with the operation of the rupture disc.

A heat shield may consist of overlapping stainless steel members that permit the pressure to build up on both sides of the members but move back out of the way when the rupture disc bursts.

Another design (see Figure 6-2) consists of a holder which is filled, in this case, with a loose wool of amorphous silica filaments. The wool is expelled from the holder upon bursting of the rupture disc.

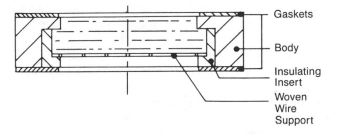

Figure 6-2. Heat shield consisting of holder filled with loose wool of amorphous silica filaments. (Courtesy of Marston Palmer Limited.)

Pressure Tolerances

Manufacturing range. The manufacturing range is an allowable range of pressures around the specified bursting pressure within which a rupture disc can be rated/stamped. This pressure range must be agreed upon between user and manufacturer. The purpose of the agreement is to permit the economical production of some types of rupture discs.

The types of rupture discs mainly involved are prebulged rupture discs made of ductile materials which by their design fail in tension. Because the tensile strength of ductile materials used in the manufacture of rupture discs is fairly high, such discs must be made of relatively thin foils, particularly for low bursting pressures. Thin foils of uniform thickness and mechanical strength, however, are difficult to produce. In addition, thickness and mechanical properties vary between heats of materials. Manufacturers must therefore select from different foils until the desired bursting pressure has been achieved. To keep manufacturing costs within acceptable limits, only a limited number of finding tests can be carried out.

The manufacturing range is expressed as a percentage of the specified bursting pressure and varies from pressure range to pressure range. If, for example, the specified bursting pressure is 10 bar (145 psi) and the manufacturing range is 11%, a disc which is rated/stamped between 8.9 and 10 bar (129 to 145 psi) meets the disc specification.

The manufacturing range may also be said to be plus 7%/minus 4%. In this case, a disc which is rated/stamped between 9.6 and 10.7 bar (139 and 155 psi) meets the disc specification.

Burst pressure tolerance. Besides the agreed manufacturing tolerances in the specified rating of some types of rupture discs, there is the burst pressure tolerance, which defines the maximum variation in bursting pressure that the disc may have from its stamped rating.

The ASME Power Boiler and Pressure Vessel Codes require the manufacturer to guarantee a burst pressure tolerance of plus/minus 5% of the stamped bursting pressure at the coincident disc temperature. This ruling does not apply to pressure vessels that have an external or internal operating pressure not exceeding 103 kPa (15 psi). In this case, the user should consult the rupture disc manufacturer on the applicable burst pressure tolerance.

Performance tolerance. The British standard 2915:1984 does not use the terms *manufacturing range* and *burst pressure tolerance* but instead uses the term *performance tolerance,* which defines the difference between the minimum and maximum specified bursting pressures at a coincident temperature. The performance tolerance therefore equals the sum of manufacturing range and burst pressure tolerance. Thus, if the manufacturing range is plus/minus 10% and the burst pressure tolerance plus/minus 5%, the equivalent performance tolerance is plus/minus 15%.

Design and Performance of Ductile Metal Rupture Discs

Prebulged rupture discs of solid construction. Prebulged rupture discs of solid construction are formed from flat discs by applying a pressure to the underside of the disc of normally above 70% of the bursting pressure. This method of manufacture gives the rupture disc the hemispherical shape shown in Figure 6-3. When the operating pressure grows beyond the predoming

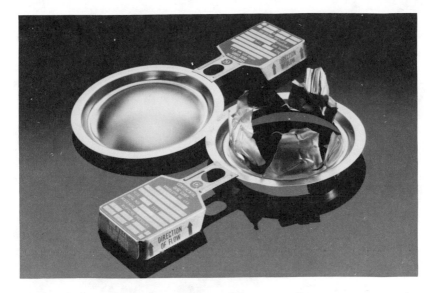

Figure 6-3. Plain solid prebulged rupture disc, before and after bursting. (Courtesy of Continental Disc Corporation.)

pressure, the dome of the disc starts to grow. As the operating pressure approaches 95% of the bursting pressure, localized thinning in the region of the dome center occurs, leading to failure as shown in Figure 6-3. This failure is accompanied by some fragmentation of the disc.

To avoid early failure of the disc due to creep, the normal operating pressure should not exceed 70% of the bursting pressure. This working pressure to bursting pressure value should be lowered still further where the operating pressure pulsates, depending on the frequency and the magnitude of the pulsating pressure. For this reason, the useful working life of this type of rupture disc cannot be predicted but must be established under working conditions. In general, plain prebulged rupture discs of solid construction are not well suited for pulsating pressure and elevated temperature conditions.

Because plain prebulged rupture discs are made of fairly thin foils, periods of vacuum or back pressure can cause the disc to crinkle or fail altogether. Where this can happen, the discs must be provided with vacuum or back pressure supports. These must be able to withstand full vacuum pressure or, in special cases, higher back pressure. Figure 6-4 shows a support for vacuum conditions. In higher back pressure applications, the disc may be provided with an additional auxiliary support, as shown in Figure 6-5.

The supports must fit the dome of the rupture disc as perfectly as possible. Should the operating pressure occasionally reach values higher than the predoming pressure, the disc will stretch and lift off the support. Vacuum condi-

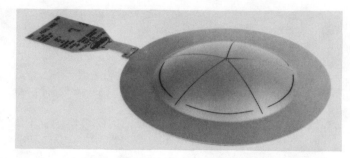

Figure 6-4. Vacuum support for prebulged rupture disc, full opening. (Courtesy of Marston Palmer Limited.)

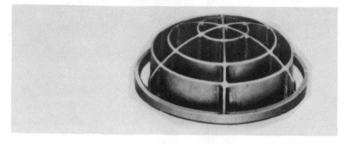

Figure 6-5. Auxiliary back pressure support to supplement vacuum support for high back pressures. (Courtesy of Marston Palmer Limited.)

tions or back pressure will then collapse the rupture disc against the support and produce a wrinkle pattern, or turtle backing, as shown in Figure 6-6. This deformation shortens the life of the disc considerably.

The manufacturing range of plain prebulged rupture discs is fairly high, and may vary between 9% and 80%. The range is highest for low pressure discs and decreases as the pressure increases. The greatest potential of these rupture discs is therefore in the higher pressure range.

To permit prebulged rupture discs to be made of thicker materials, the external face of such discs has been provided with cross scores, as shown in Figure 6-7. These scores represent weak lines along which the rupture disc bursts.

Cross-scored rupture discs offer a number of advantages over plain discs. First, the maximum operating pressure can be raised to 80% of the bursting pressure. Second, close control of the score depth permits the manufacturing range to be controlled within 10%, and in special cases to 5%. Third, the disc is designed to open without fragmentation and, therefore, may be used under pressure relief valves. Under vacuum and back pressure conditions, the scored disc requires vacuum or back pressure support like plain discs.

Both types of prebulged rupture discs may be used in gas and liquid service, including service in totally full liquid systems.

Prebulged composite rupture discs. Prebulged composite rupture discs consist of two or more members to control the bursting pressure, as in the rupture discs shown in Figures 6-8 and 6-9. The top member represents a prebulged metal rupture disc, similar to the solid prebulged rupture disc, but slit in a predetermined configuration. The purpose of the slits is to weaken the disc so that bursting occurs at lower pressures. A second member, generally made of plastic, and in some cases of metal such as lead, silver, or gold, fits to the concave side of the slit disc and provides the fluid seal. The com-

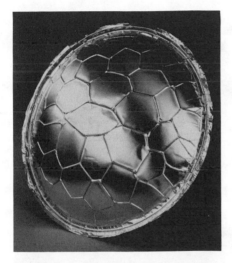

Figure 6-6. Wrinkle pattern or turtle backing of prebulged rupture disc as a result of reverse flexing against inadequate support or stretching due to pressure excursion and then reverse flexing. (Courtesy of Continental Disc Corporation.)

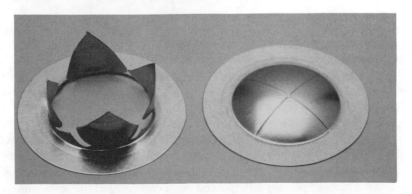

Figure 6-7. Cross-scored prebulged rupture disc before and after bursting. (Courtesy of Continental Disc Corporation.)

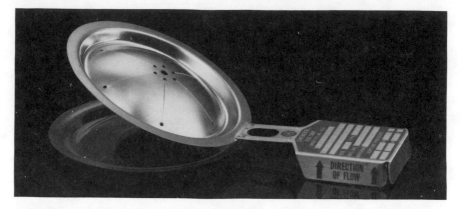

Figure 6-8. Prebulged composite rupture disc, consisting of slotted top member, followed by a seal member. (Courtesy of Continental Disc Corporation.)

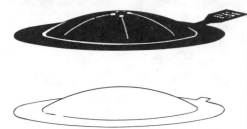

Figure 6-9. Exploded view of pre-bulged composite rupture disc, showing slotted top section followed by seal member and vacuum support. (Courtesy of Marston Palmer Limited.)

bined action of both members determines the bursting pressure. If vacuum or back pressure can occur, a vacuum or back pressure support becomes the third member of the assembly.

Prebulged composite rupture discs have a number of advantages over solid prebulged rupture discs:

- They can be designed to burst at a considerably lower pressure than solid prebulged rupture discs. For example, the lowest bursting pressure for which a size DN 50 (NPS 2) solid prebulged rupture disc from the thinnest available foil can be made is about 10 bar (150 psi), compared with about 1 bar (15 psi) for a prebulged composite disc of the same material.
- The allowable operating pressure of these discs can be raised to 80% of the bursting pressure.

- The choice of linings can considerably improve the corrosion resistance of the top member. The top member may therefore often be made of a less expensive material than otherwise required.
- The top member may be designed to prevent metal fragmentation.

On the debit side, prebulged composite rupture discs are suitable only for the lower pressure range, unlike solid prebulged rupture discs.

Figure 6-10 shows three different slot-and-hole patterns used in prebulged composite rupture discs.

The first pattern consists of a six-hole center section and six slots with pierced holes at each end. The distance between the holes, and the thickness and strength of the material control the bursting pressure. Fracture at the time of burst occurs in the center section from hole to hole in a hexagon shape. The center section can completely fragment off.

	Failure Mode	Advantages	Potential Problems
Six Hole Pattern	 Break occurs around the holes, leaving the center section to be held by one pie-shaped section.	– Low burst pressure – 80% of disc burst pressure – Burst accuracy ± 5%	– Center section may be completely ripped off as one piece. – Center section may hinder flow path: – Metal seal or metal liner may fragment
Seven Hole Pattern	 Disc breaks radially into six wedge-shaped sections.	– Low burst pressure – 80% of disc burst pressure – Non-fragmenting design using plastic or Teflon seal – Burst accuracy ± 5% – Teflon seal protects disc from fluid	– Very small pieces of center section may fragment when supplied with Teflon liner only. – Metal seals or metal liner may fragment
Toilet Seat	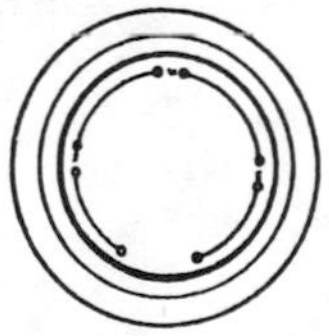 Disc breaks as one piece.	– Low burst pressure – % of burst pressure varies dependent on pressure range – Tolerance-accuracy dependent on burst range	– Center section may completely fragment out. – Center section blocks flow path – Metal seals or metal liner may fragment

Figure 6-10. Hole and slot pattern in prebulged composite rupture discs. (Courtesy of Continental Disc Corporation.)

The second pattern is similar to the first but includes a center hole which causes the disc to rupture radially. Rupture discs with this hole pattern do not tend to fragment when PTFE seals are used. At high flow rates or higher pressures, however, small fragments of the PTFE seal may still tear off. However, these fragments are usually so small that they do not tend to affect the downstream system. Metal seals will fragment.

The third pattern is commonly referred to as the toilet seat pattern because of its mode of opening in the manner of a toilet seat. This pattern is made up of four circumferential slot sections with one large and three smaller unslotted sections. Failure of the disc occurs along the smaller unslotted sections, while the longer unslotted section keeps the middle section from fragmenting into the flow path.

The application of prebulged composite rupture discs covers gas and liquid service in a manner similar to solid prebulged rupture discs.

Flat composite rupture discs. Flat composite rupture discs such as that shown in Figure 6-11 are designed for low pressure applications only, but are otherwise identical in design to their prebulged counterpart. They may be used for failure in one direction only, in which case the slotted disc is lined with a seal member on one side only; or for failure in both directions, in which case the slotted disc is lined on both sides.

The discs are generally offered in sizes from DN 50 (NPS 2) through DN 300 (NPS 12) for bursting pressures of 0.07 bar (1 psi) to 1 bar (15 psi), depending on size and operating temperature. The discs are frequently used for the protection of atmospheric vessels.

Figure 6-11. Flat composite rupture disc. (Courtesy of Continental Disc Corporation.)

This particular rupture disc does not require a holder and may be bolted between standard pipe flanges. The operating pressure is not recommended to exceed 50% of the rated bursting pressure of the disc.

Reverse buckling discs. The continuing effort to reduce the operating margin and manufacturing range of rupture discs, and also to obtain non-fragmenting discs, led to the development of reverse buckling discs.

Contrary to prebulged rupture discs, which fail in tension, reverse buckling discs fail in compression, causing the disc to buckle on failure. This buckling occurs at a stress level considerably lower than for prebulged rupture discs.

Reverse buckling discs must therefore be made thicker than prebulged rupture discs, in many cases about 6 times thicker for the same bursting pressure.

Also, the material property which determines the buckling strength is the *Young's modulus*. This property is more constant and more easily reproducible than the ultimate tensile strength. Consequently, reverse buckling discs are much easier to manufacture to close tolerances.

The pressure at which the disc will buckle is determined not only by the properties of the material, but also by the shape of the dome and of the supporting edges of the disc holder. When assembling disc and holder, great care must therefore be taken that disc and holder are perfectly matched. There must be no dirt between disc and holder. Discs and holders with even small damage must be avoided.

Unlike prebulged rupture discs, reverse buckling discs do not creep or stretch as pressure builds up. When the operating pressure reaches the bursting pressure, the reverse buckling disc snaps through the neutral position into a completely opposite configuration. By itself, the disc does not burst open on reversal. This is achieved by either a cutting device against which the disc must be slammed, or scoring the disc. Alternatively, the disc may be designed to slip out of the holder upon reversal without bursting.

Figure 6-12 shows a reverse buckling disc which relies on knife blades to cut the disc open. However, the disc must strike the knife blades with high energy for this to happen. For this reason, the disc may be used only in gas service, and in liquid service in which there exists a substantial volume of gas between the liquid and the disc. In totally full liquid systems, however, where the speed of disc reversal may be slow, the disc may initially come to rest on the blades and then be cut open only after the system pressure has increased substantially, frequently exceeding the burst pressure of the system.

It is essential also that the edges of the cutting device are kept sharp. If this is neglected, the installation can become unsafe.[75] The cutting edges must, therefore, be checked on a regular basis and be resharpened if necessary.

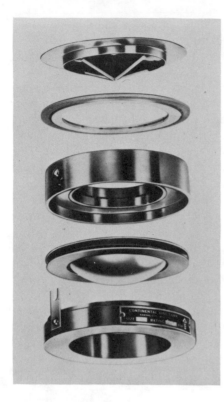

Figure 6-12. Exploded view of rupture disc device, incorporating reverse buckling disc with knife blades. (Courtesy of Continental Disc Corporation.)

The manufacturer should be contacted for advice in this case. Care must be taken not to change the blade location or configuration. In most cases, the manufacturer should perform repair or replacement.

Further development led to the design of reverse buckling discs with score lines. These lines represent weak lines along which the disc bursts open after reversal. Two line patterns are used, one cross score pattern which divides the disc into pie shaped sections, and one which forms a partial circle around the dome along the seat of the disc.

Cross-scored reverse acting discs will break open upon reversal into pie shaped sections, with the base firmly held by the holder. When used in gas service, and in liquid service in which there is a substantial volume of gas between the liquid and the disc, the expanding gas will open the disc in milliseconds. However, the disc may not be used in totally full liquid systems.

In the case of the disc with partial circumferential score lines shown in Figure 6-13, the disc fractures upon reversal along the score line, with the fracture migrating around the disc. The pressure differential across the disc then swings the center portion around the unscored section, as shown in Figure 6-14. When used in gas service, the disc opens in milliseconds. When used in totally full liquid systems, the fluid acting on the large undivided area of the center portion of the disc forces the disc fully open with little rise in overpressure. The disc may therefore be used not only in gas service, but also in totally full liquid systems.

The disc shown in Figure 6-15 is unique in that it does not burst upon reversal but slips out of the holder in one piece, so this mode of opening offers the maximum venting capacity. Because venting of the disc does not rely on the assistance of either knife blades or grooving, the disc can be offered for high pressure applications.

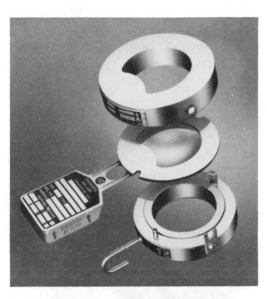

Figure 6-13. Exploded view of rupture disc device incorporating reverse buckling disc with partial circumferential score line. (Courtesy of Continental Disc Corporation.)

Figure 6-14. Burst reverse buckling discs with partial circumferential score lines. (Courtesy of Continental Disc Corporation.)

When used in gas service, and in liquid systems in which there is a substantial gas volume between the disc and the liquid, the disc is, upon reversal, expelled from the disc holder in milliseconds. However, the disc may not be used in totally full liquid systems except in the larger sizes. If the use of this disc is contemplated for this service, the manufacturer must be consulted.

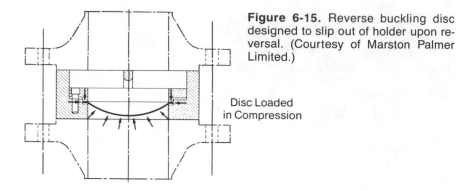

Figure 6-15. Reverse buckling disc designed to slip out of holder upon reversal. (Courtesy of Marston Palmer Limited.)

Disc Loaded
in Compression

Vacuum or back pressure puts reverse buckling discs into tension under which they withstand a higher pressure than under forward pressure. For scored reverse buckling discs, the ratio of back pressure to forward pressure capability is generally 1.5 or less. For unscored reverse buckling discs, the back pressure capability is considerably higher than the forward pressure capability. An exception is the case of reverse buckling discs which slip out of the holder upon reversal (see Figure 6-15). For these, the ratio of back pressure to forward pressure capability is generally less than 1.0, with a maximum of 1.1 for small sizes.

This ratio of back pressure to forward pressure capability has two consequences. First, reverse buckling discs do not generally require vacuum or back pressure supports. Second, the disc must be installed in the correct direction to prevent overpressurizing the pressure system. This overpressurizing due to wrong installation is particularly serious for unscored reverse buckling discs, except for the type shown in Figure 6-15 which is expelled from the holder upon reversal. With scored reverse buckling discs, the pressure in the pressure system could rise to 1.5 times the maximum allowable working pressure, which in most installations would equal the test pressure of the system.

To overcome the problem of wrong installation with these discs, manufacturers have developed holders which permit installation of the disc in one direction only.

Because reverse buckling discs operate at low stress levels, they will not deform permanently until just below the bursting pressure. They may therefore be operated at pressures up to 90% of the bursting pressure. Furthermore, the discs will tolerate pressure fluctuations up to 90% of the bursting pressure without fatigue failure.

Reverse buckling discs are offered with zero or narrow manufacturing range and relatively small burst pressure tolerances.

Compared with prebulged rupture discs, reverse buckling discs are much more sophisticated. They are more accurate and permit operation much closer to the bursting pressure. On the debit side, they are more expensive than prebulged rupture discs and require more care in selection and installation. Reverse buckling discs with knife blades also need care in maintaining the knife blades.

Ultra low pressure rupture discs.
The disc shown in Figure 6-16 is an example of an ultra low pressure rupture disc which operates in both positive pressure and vacuum directions. The construction can be described as a combination of a reverse buckling disc with a composite prebulged rupture disc.

The disc consists of a perforated metal section that uses the center seven-hole configuration of the composite prebulged rupture disc, followed by a PTFE seal, a support girdle, and a set of knife blades, as shown. The perforations in the metal section provide the relief area when pressure conditions force the PTFE seal against the knife blades.

If the disc is mounted with the concave side facing the medium and exposed to a positive system pressure, the disc acts as a prebulged composite rupture disc. When the pressure acts in the reverse direction, the PTFE seal is forced against the gir-

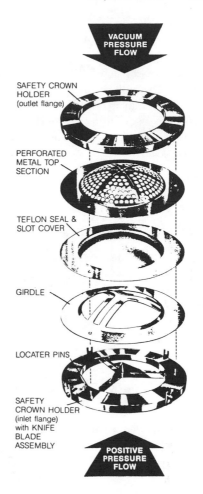

Figure 6-16. Ultra low pressure rupture disc. (Courtesy of Continental Disc Corporation.)

dle until the girdle reverses. At this point, the seal snaps through and is cut by the knife blades. Used in this manner, the disc may be designed to burst at a positive system pressure between 0.1 bar (1.5 psi) and 10 bar (150 psi) or at a negative system pressure between 25 mm (1 in.) and 750 mm (30 in.) W.G.

If the disc is mounted with the convex side facing the medium, the action of the disc is reversed. Under these conditions the disc may be designed to burst at a positive system pressure between 25 mm (1 in.) and 750 mm (30 in.) W.G., or at a negative system pressure or superimposed back pressure between 0.1 bar (1.5 psi) and 10 bar (150 psi).

Vent panels. Vent panels are large-area low pressure relieving devices for closed spaces in which there is the possibility of dust, mist or gas deflagration. The shape of these panels may be either circular, rectangular or square.

Figures 6-17 and 6-18 show examples of such panels. In these particular cases, the vent panels are flat composite constructions, consisting of a slotted top section that is backed by a liner representing the seal. In a deflagration, the panels open along the slotted lines to relieve the peak pressure.

Other vent panel constructions may use scored metal panels, or slit and taped metal panels.

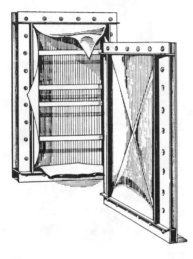

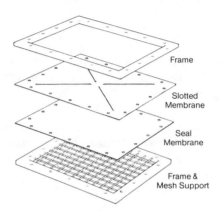

Figure 6-17. Vent panel of composite construction before and after bursting. (Courtesy of Continental Disc Corporation.)

Figure 6-18. Exploded view of vent panel of composite construction. (Courtesy of Marston Palmer Limited.)

Design and Performance of Graphite Rupture Discs

Graphite is a valuable material from which low pressure rupture discs can be produced. The material usually performs well in most corrosive environments.

Two grades of graphite are used in the manufacture of rupture discs. One grade is porous and is sealed by impregnation with phenolic or furane resins. The other grade is pure graphite, which in compressed form is impervious to fluids. Depending on the grade of graphite used, different types of disc constructions have evolved.

Resin impregnated graphite rupture discs. Graphite discs were originally made in monoblock form only. These are one-piece devices that combine a flat bursting membrane with the mounting flange, as shown in Figure 6-19.

Such discs are robust and tend to perform accurately if properly installed. However, overtightening can cause the graphite to crack. To overcome this problem, the outside diameter of the graphite disc may be surrounded by a metal ring, as shown in Figure 6-20.

To achieve greater economy in the use of graphite, some manufacturers have standardized designs in which the bursting membrane is a separate unit from the flange and is replaceable. The disc is held in this case in a two-part holder with gaskets on both sides of the disc. It is essential, however, that the graphite not be overstressed by overtightening the flange bolts. This has been achieved in a design shown in Figure 6-21. The pressure side of the

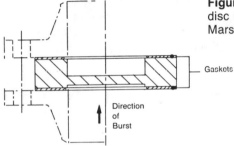

Figure 6-19. Monoblock type rupture disc of brittle graphite. (Courtesy of Marston Palmer Limited.)

Gaskets

Direction of Burst

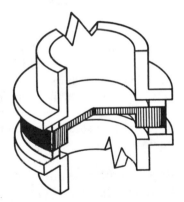

Figure 6-20. Rupture disc of brittle graphite, reinforced on its circumference by a metal ring. (Courtesy of Continental Disc Corporation.)

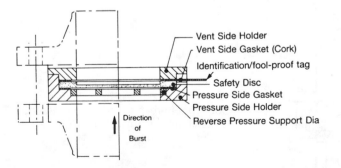

Vent Side Holder
Vent Side Gasket (Cork)
Identification/fool-proof tag
Safety Disc
Pressure Side Gasket
Pressure Side Holder
Reverse Pressure Support Dia

Direction of Burst

Figure 6-21. Rupture disc device with replaceable disc of brittle graphite, non-torque-sensitive construction. (Courtesy of Marston Palmer Limited.)

holder contains a recess which receives both the graphite disc with gasket and the spigot of the clamping ring, representing the vent side of the holder. When the two parts of the holder are clamped firmly together, the distance between the clamping faces is smaller than the overall thickness of the graphite disc with gaskets by a controlled amount, so as not to overstress the graphite disc.

Such discs will withstand vacuum or back pressure equal to the normal operating pressure. If the back pressure can rise above the normal operating pressure, a back pressure support must be provided such as that shown in Figure 6-21. However, the support may considerably reduce the available vent area.

The graphite rupture disc shown in Figure 6-22 is designed to provide protection against two different pressures from opposite directions. The disc consists of two components: a rupture disc of monoblock construction, and a perforated disc placed on the vent side and bonded on its circumference to the monoblock disc. A typical application of this disc is the protection of closed storage tanks against malfunctioning of primary breathers during loading and unloading.

The graphite rupture discs are offered with zero manufacturing range and a burst pressure tolerance of plus/minus 5% for a bursting pressure of 1 bar (14.5 psi) and above, and of plus/minus 0.05 bar (0.75 psi) for bursting pressures below 1 bar (14.5 psi). However, the operating pressure should not exceed 75% of the bursting pressure.

In contrast to ductile metal rupture discs, graphite rupture discs give full bore opening upon bursting irrespective of the type of service. The discs therefore give safe protection not only in gas service, but also in totally full liquid systems.

Pure graphite rupture discs. Pure graphite discs are produced in domed form with the center domed portion surrounded by a flange. The flange is

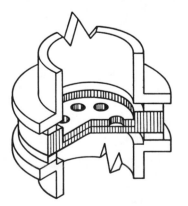

Figure 6-22. Rupture disc of brittle graphite for two different pressures from opposite directions. (Courtesy of Continental Disc Corporation.)

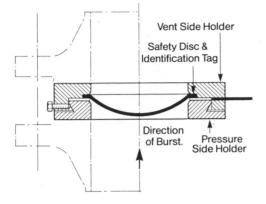

Vent Side Holder

Safety Disc &
Identification Tag

Direction
of Burst.

Pressure
Side Holder

Figure 6-23. Rupture disc device with reverse buckling disc of pure graphite. (Courtesy of Marston Palmer Limited.)

required to support the disc when gripped in a holder. They are normally installed as reverse buckling discs only, as in the installation shown in Figure 6-23.

Because reverse buckling discs operate in a compression mode, creep and fatigue are minimal. The maximum permissible operating pressure in percent of the bursting pressure is high, being 90% over a temperature range of 20°C (68°F) to 650°C (1200°F). A principal advantage of the disc is low burst pressure combined with narrow burst tolerances.

Extensive pressure cycling tests carried out at 20°C (68°F) have proven the creep and fatigue resistance of these discs. Discs have been pressure loaded between 0–90% of the bursting pressure at a rate of 30 cycles/minute for 90,000 cycles without failure of the discs.

Pure graphite offers a corrosion resistance similar to impregnated graphite discs. However, care must be taken that pure graphite discs are not exposed to erosion.

Like impregnated rupture discs, pure graphite rupture discs give full bore opening upon bursting irrespective of the type of fluid. To catch the spent disc, a suitable disc arrester may be required.

The following performance data of pure graphite reverse buckling discs have been quoted:*

Size: DN	25	50	80	100	150
NPS	1	2	3	4	6

Offered bursting pressures

Minimum bar	0.4	0.14	0.10	0.04	0.04
psi	5.8	2.0	1.4	0.6	0.6

*By courtesy of Marston Palmer Limited

| Maximum bar | 15 | 10 | 4.5 | 2.5 | 2.5 |
| psi | 218 | 145 | 65 | 36 | 22 |

Burst pressure tolerance plus/minus %

| lowest at below 0.5 bar (7.2 psi) | 10 | 15 | 15 | 20 | 15 |
| at 0.5 bar (7.2 psi) and higher | 5 | 5 | 5 | 5 | 5 |

Ratio of working pressure/bursting pressure = 0.9 at 20°C (68°F)
through 600°C (1110°F)

Maximum operating temperatures: 350°C (660°F) oxidizing atmosphere
3000°C (5400°F) inert atmosphere

Pure graphite discs may also be combined with metal reverse buckling discs, as shown in Figure 6-24. This combination represents a pure-graphite lined reverse-buckling disc of the type shown in Figure 6-15 in which the metal disc provides the bulk of the strength. A particular field of application of this combination is in PVC reactors, because of the non-stick properties of graphite to PVC.

Figure 6-25 shows a variation of this combination in which the metal disc is perforated. This design allows this combination to be used for overpressure protection as well as vacuum or back pressure protection. The hole diameters in the metal disc are chosen to give adequate support to the graphite disc. The number of holes is related to the required relief capacity under vacuum or back pressure conditions.

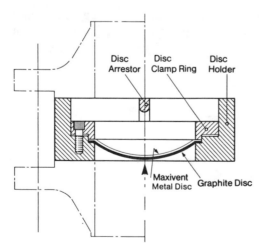

Figure 6-24. Rupture disc device with ductile metal rupture disc similar to Figure 6-15, with pure graphite rupture disc backing. (Courtesy of Marston Palmer Limited.)

Rupture Disc Holders

The original disc holders were in-tended for metal rupture discs which failed in tension. They consisted of two flat faced pipe flanges between which the discs were clamped.

The early designs encountered two problems. First, the disc would tend to shear on the sharp edge of the top flange, resulting in unpredictable failure. Second, as pressure in-creased, the disc would tend to slip between the flat faces, resulting in leakage and unpredictable failure.

The first problem was overcome by radiusing the inside corner of the top flange. To overcome the second problem, designers changed the flat seat to a 30° conical seat which wedges the disc between two faces, as shown in Figure 6-26.

This configuration also has a debit side. Overtightening of the conical seat can cause the very thin foils to thin at the base of the seat, leading to early rupture disc failure. Over-tightening can also deform the coni-cal lip of the outlet holder part, al-though this has been overcome for high pressure applications by provid-ing a heavy conical lip. For best per-formance of the rupture disc, the torque recommendations of the man-ufacturer must be observed.

Metal Disc

Graphite Disc

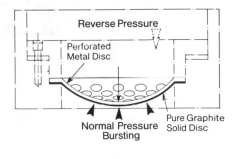

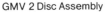

Reverse Pressure

Perforated Metal Disc

Normal Pressure Bursting

Pure Graphite Solid Disc

GMV 2 Disc Assembly

Figure 6-25. Rupture disc device for two-way protection, employing perfo-rated ductile metal reverse buckling disc similar to Figure 6-15, with pure graphite rupture disc backing. (Cour-tesy of Marston Palmer Limited.)

The 30° conical seat configuration is standard in the United States. How-ever, not all manufacturers have common interchangeable seat configura-tions. In many parts of Europe and the Far East, the flat seat has been re-tained for the lower pressure applications because of the advantage of not thinning the very thin foils at the base of the conical seat.

Reverse buckling discs, which by their design fail in compression, do not present the problem of slippage between the flanges, so the seating faces can be flat in this case irrespective of pressure. It is essential, however, that the

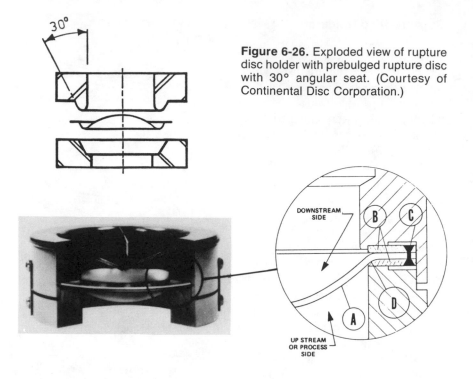

Figure 6-26. Exploded view of rupture disc holder with prebulged rupture disc with 30° angular seat. (Courtesy of Continental Disc Corporation.)

Figure 6-27. Scrap view of rupture disc holder for reverse buckling disc with capturing rings designed for foolproof erection of rupture disc. (Courtesy of Continental Disc Corporation.)

supporting edges of the holder do not deform the dome or the edges of the dome. Any such deformation would affect the bursting pressure.

Manufacturers have developed a number of holder designs that ensure foolproof erection of reverse buckling discs, as in the holders shown in Figures 6-13 and 6-27.

In the design in Figure 6-27, the collar of the disc is combined with capturing rings which provide the correctly shaped supporting edges. Both the capturing ring and the inlet holder flange are then centered in a recess in the outlet holder flange.

In the design in Figure 6-13, the disc is centered by two pins on the inlet flange of the disc holder in a manner which permits the disc to be pointing in one direction only. To also ensure mounting of the assembly in the correct direction, the inlet flange of the holder is provided with a permanent J-bolt. This bolt must match a hole in the rim of the flange to which the rupture disc device is to be mounted. A three-dimensional tag of the rupture disc provides immediate visual verification of the correctness of the installation.

Disc holders for graphite discs are designed to prevent the disc from being crushed. Figures 6-21, 6-23, and 6-24 show some of the holders designed for this purpose.

Clean-Sweep Assembly

The clean-sweep assembly shown in Figure 6-28 is used in pipelines in which the flowing fluid has a tendency to build up solids under the disc if the disc is mounted on a dead pipe branch. The solids build-up could prevent the rupture disc from protecting the system from overpressure conditions.

The clean-sweep design eliminates the dead pocket and enables the product to sweep across the rupture disc surface. This action tends to keep the disc clear from solids build-up and to assure proper functioning of the disc.

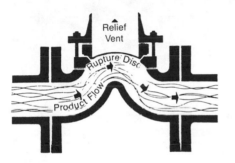

Figure 6-28. Clean-sweep assembly of rupture disc device and pipeline section, incorporating prebulged rupture disc. (Courtesy of Continental Disc Corporation.)

Quick-Change Housings

To permit rapid replacement of rupture discs after bursting, a number of quick-change housings for the rupture disc device have been developed, such as those shown in Figures 6-29 and 6-30.

These holders require the rupture disc device to slide sideways into the vent system. To achieve a fluid seal between the rupture disc device and the vent system, the inlet and outlet flanges of the rupture disc device are provided with O-ring seals which are carried in dove-tail grooves. When the O-rings are put under fluid pressure, they will key into the irregularities of the mating flange face and lock the rupture disc device firmly into position.

Design of Double Disc Assemblies

Double disc assemblies consist of two rupture discs, which may be held in a common holder, as shown in Figure 6-31, or in two separate holders with a pipe section in between.

Figure 6-29. Quick-change housing. (Courtesy of Continental Disc Corporation.)

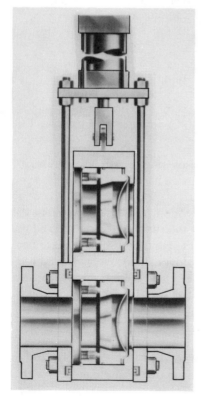

Figure 6-30. Quick-change housing, sliding gate construction. (Courtesy of Continental Disc Corporation.)

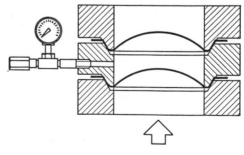

Figure 6-31. Rupture disc device, containing two rupture discs in series. (Courtesy of Sempell Armaturen.)

The requirements for the installation of such assemblies for pressure relief purposes are contained in the British standard 2915:1985 (C5). Accordingly, the space between the rupture discs should be large enough to ensure that the correct functioning of the rupture disc is not impaired. Also, the space between the rupture discs should be provided with a pressure monitoring device or be vented by means of an excess flow valve.

The ASME code does not deal with double disc assemblies.

The purpose of such installation is to prevent loss of product due to premature failure of the first disc from fatigue or corrosion. However, if the first disc bursts at a pressure close to the bursting pressure of the second disc, the sudden pressure shock may also burst the second disc.

A double disc assembly may also be used as a quick opening valve. Such assemblies may contain two reverse buckling discs which may be designed, for example, to burst at a pressure equal to 60% of the system pressure. To prevent the discs from bursting prematurely, the space between the discs can be pressurized to 50% of the system pressure. When releasing this pressure, the discs will open within a few milliseconds.

Design of Rupture Disc and Pressure Relief Valve Combinations

Rupture discs may be combined with pressure relief valves in three ways, namely by mounting the rupture disc either to the inlet side or the outlet side of the pressure relief valve, or by mounting one rupture disc each to the inlet and the outlet side.

In most installations of rupture disc and pressure relief valve combinations, the rupture disc is mounted to the inlet side of the pressure relief valve. Only occasionally is the rupture disc mounted to the outlet side and still less frequently to the inlet and the outlet side of the pressure relief valve.

The ASME Code, Section VIII, Division 1 addresses the design requirements for mounting the rupture disc to either the inlet or the outlet side of the pressure relief valve, but does not mention the installation of one rupture disc each to the inlet and the outlet side of the pressure relief valve. The British standard 2915:1984, appendix C, on the other hand, covers the design requirements for all three types of installation.

The following information covers the ASME Code requirements. However, this information is given as reference only. In all cases, the user should follow the governing regulations which apply to his installation.

Rupture disc mounted to the inlet side of a pressure relief valve. The primary use for a rupture disc mounted to the inlet side of a pressure relief valve is to protect the pressure relief valve from corrosive media and to minimize or eliminate valve leakage.

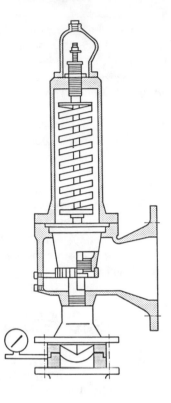

Figure 6-32. Rupture disc mounted to the inlet side of pressure relief valve. (Courtesy of Sempell Armaturen.)

The rupture disc in such a combination is shown in Figure 6-32. In general practice, the set point of the rupture disc should be as close to the set pressure of the pressure relief valve as practical. There are no guidelines, rules, or regulations presently established for this relationship. In no case should the rupture disc set point exceed the maximum allowable working pressure of the vessel.

The rupture disc should also be of a non-fragmenting design. This type of disc prevents fragments from interfering with the operation of the pressure relief valve.

The ASME Boiler and Pressure Vessel Code, Section VIII, UG-127, dealing with non-reclosing pressure relief devices, stipulates the following for pressure relief valve/rupture disc assemblies:

- The word *combination* is used to designate a pressure relief valve/rupture disc assembly when the rupture disc is installed between the pressure relief valve and the vessel. Ref: UG-127 (a) (3) (b).
- The pressure relief valve must meet the capacity requirements of UG-133 (a) and (b). Ref: UG-127 (a) (3) (b) (1).
- An untested combination of rupture disc and pressure relief valve must be derated by 20% of the stamped capacity of the pressure relief valve; i.e., the capacity of the untested combination is 0.8 times the stamped capacity of the pressure relief valve. Ref: UG-127 (a) (3) (b) (2).
- As an option, the pressure relief valve and rupture disc combination may be tested to UG-132 to determine the actual combination capacity factor. This certified combination capacity factor may be used in place of the 0.8 multiplier. Ref: UG-127 (a) (3) (b) (3).
- The space between the rupture disc and the pressure relief valve must be free vented or have a suitable pressure indicating device and bleed device to detect leakage of the disc and to indicate pressure buildup be-

tween the disc and the pressure relief valve. This indicator would also be used to indicate if a disc has blown. Ref: UG-127 (a) (3) (b) (4).
- The rupture disc device opening, after it has burst, is to provide sufficient flow equal to the capacity of the pressure relief valve. Ref: UG-127 (a) (3) (b) (5).

Rupture disc mounted to the outlet of a pressure relief valve. The use of a rupture disc on the outlet side of a pressure relief valve is permitted by the ASME Code. The disc is used on the outlet to protect the pressure relief valve from hazardous materials that may be present in the header system or the environment. Rupture discs are also used when a pressure relief valve is mounted to a common header system. In this case, the rupture disc prevents back pressure and contamination from reaching the pressure relief valve. Back pressure and contamination would be caused during an upset in another part of the connecting system.

The rupture disc used in this assembly may also be of the fragmenting type, provided the downstream system tolerates the ingress of rupture disc fragments or the fragments are being caught.

According to the ASME Code UG-127 (a) (3) (c), the following is stipulated:

- The pressure relief valve is to be designed to open at the proper set pressure regardless of any back pressure between the rupture disc and the pressure relief valve. Ref: UG-127 (a) (3) (c) (1).
- The space between the rupture disc and the pressure relief valve shall be vented or drained to prevent accumulation of pressure due to leakage of the pressure relief valve. Ref: UG-127 (a) (3) (c) (1).
- The pressure relief valve must meet the capacity requirements of UG-133 (a) and (b). Ref: UG-127 (a) (3) (c) (2).
- The opening through the rupture disc after burst is to provide sufficient flow equal to the rated capacity of the pressure relief valve. Ref: UG-127 (a) (3) (c) (4).
- The stamped/rated bursting pressure of the rupture disc plus any superimposed back pressure shall not exceed:
    1. The maximum allowable working pressure of the vessel.
    2. The design pressure of the outlet side of the pressure relief valve and of the outlet piping.
    3. The set pressure of the pressure relief valve.

Notes:

- To meet the requirements of UG-127 (a) (3) (c) (1), the pressure relief valve must be of the balanced type.

- The ASME Code does not stipulate a derating of the pressure relief valve capacity under section UG-127. However, if built-up back pressure resulting from flow through the rupture disc turns sonic flow through the pressure relief valve into subsonic flow, this must be taken into account when assessing the flow capacity of the pressure relief valve/rupture disc assembly.

Rupture discs mounted to the inlet and outlet side of a pressure relief valve. To design an installation using rupture discs on both the inlet and outlet of a pressure relief valve, the code requirements for both rupture disc locations must be met.

Sizing of Rupture Discs

The ASME Code, Section VIII, Division 1, and BS 2915:1984 present the equations for sizing rupture discs in gas service. In these equations it is assumed that the gas follows the ideal gas laws and that flow through the rupture disc is isentropic. It is also assumed that the upstream flow velocity is small compared with the flow velocity through the rupture disc and, therefore, can be neglected. The equations are similar to those for pressure relief valves, as presented in Chapter 5.

The following are intended for reference purposes:

- The flow area in the sizing equation shall be the minimum net area existing after disc burst. Ref: UG-127 (a) (2) (a) and associated footnote 37.
- The applicable coefficient of discharge in the sizing equation shall be 0.62, which is the coefficient of discharge of a sharp edged orifice discharging to atmosphere. Ref: UG-127 (a) (2) (a), BS 2915:1984 (E1).
- The sizing equations should not be used:
    1. Where there is significant frictional pressure drop in the line before the rupture disc.
    2. Where there is significant frictional pressure drop in the line after the rupture disc.
    3. For explosive reactions or where reactions continue in the pipeline. Ref: BS 2915:1984 (E1).
    4. Where pressure loss in the vent system gives a flow rate different from that obtained with a coefficient of discharge of 0.62, the discharge capacity should be calculated using equations involving the resistance coefficient ζ. Ref: BS 2915:1984 (E1). Note: Vent pipes attached to rupture discs diminish or can severely diminish the flow capacity of rupture discs. To prevent the flow capacity in such installations from falling below the required capacity, the size of the rupture must be increased and/or the size of the vent pipe be

increased, possibly in conjunction with mounting the rupture disc to a converging nozzle.

5. Vents for the release of overpressure resulting from the deflagration of dust, gases, or mists in enclosures may be sized in accordance to guidelines of which the following are frequently accepted:

ANSI/NFPA: *Explosion Venting,* published by National Fire Protection Association, Inc.

VDI 3673:   *Pressure Release of Dust Explosions,* published by VDI-Verlag, Duesseldorf (West Germany).

Set pressure and overpressure limits in compliance with ASME code. The ASME Code, Section VIII, stipulates the permissible set pressures and overpressures at which pressure relief devices may operate. The magnitude of these pressures depends on application and source of overpressure:

1. **Single pressure relief device, non-fire:** All pressure vessels, other than unfired steam boilers, shall be protected by a pressure relief device that shall prevent the pressure from rising more than 10% above the maximum allowable working pressure (MAWP). Ref: UG-125 (c). The set pressure of these devices shall be at or lower than the MAWP. Ref.: UG-134.
2. **Multiple pressure relief devices, non-fire:** When multiple pressure relief devices are provided, they shall prevent the pressure from rising more than 16% above the MAWP. Ref: UG-125 (c) (1). In this case, one pressure relief device must be set at or below the MAWP, while the remaining pressure relief devices may be set at or below 105% of the MAWP. Ref.: UG-134 (a).
3. **Pressure relief devices—non-fire and fire conditions:** When pressure relief devices installed for non-fire conditions are exposed to fire conditions, the operating pressure under these conditions shall not rise more than 21% above the MAWP. If this overpressure limit cannot be ensured, supplemental pressure relief devices must be installed. Ref.: UG-125 (2), UG-134 (b).
4. **Supplemental pressure relief devices for fire conditions:** Supplemental pressure relief devices for fire conditions shall prevent the pressure to rise more than 21% above the MAWP. Ref.: UG-125 (2). The set pressure shall be at or below 110% of the MAWP. Ref.: UG-134 (b).

Coefficient of discharge of double rupture disc assemblies. The coefficient of discharge for double rupture disc assemblies is customarily taken as K = 0.62, as for single rupture disc installations.

Combination capacity factor. Where a rupture disc is mounted to the inlet side of a pressure relief valve, the relief capacity of this combination is determined by multiplying the rated capacity of the pressure relief valve with a combination capacity factor.

The ASME Code UG-127 (3) (2) permits a combination capacity factor of 0.80, while BS 2915:1984, appendix C, permits a combination capacity factor of 0.90.

Alternatively, the combination capacity factor may be determined from tests in accordance with the requirements of the ASME Code or BS 2915 and be certified.

According to the ASME Code requirements, the pressure relief valve used in these capacity tests shall have the largest orifice used in the particular inlet size. This ruling leads to conservative results in applications where the orifice of the pressure relief valve used in the combination is smaller than the orifice used in the test.

Tests[74] carried out with gas using rupture discs in combination with pressure relief valves have produced combination capacity factors in the order of 0.98 to close to 1.0, although a few tests have produced combination capacity factors of less than 0.95.

Certified ASME combination capacity factors may be obtained from "Pressure Relief Device Certifications and Listings," furnished by the National Board of Boiler and Pressure Vessel Inspectors.

Resistance coefficient ζ of rupture discs. The pressure loss in the inlet piping of pressure relief valves should not exceed 3% of the system pressure. This practice is designed to minimize the possibility of valve cycling in gas service due to complete loss of blowdown, as discussed in Chapter 5. The ASME Code and API RP520 are some of the codes that have accepted this practice. To carry out the appropriate pressure loss calculation for installations in which the rupture disc is mounted to the inlet side of a pressure relief valve, the resistance coefficient of the burst rupture disc must be known.

Available test data[74] indicate that rupture discs with unobstructed flow passages display a resistance coefficient ζ in the order of 0.5 or 0.6. In the case of reverse buckling discs in which the flow passage is obstructed by knife blades, values of $\zeta = 1.0$ and higher have been reported.

Fortunately, most pressure relief valve installations in combination with a rupture disc can accommodate these resistance coefficient values. However, where this value is of concern, the manufacturer should be consulted.

Rupture Disc Burst Indicators

The bursting of rupture discs may be noted in the following ways:

- By observing pressure loss in the pressure system.
- By visual inspection.

- By the sound of the rupture disc bursting.
- By remote indication.

Two categories of indicators for the remote indication of rupture disc failure are used:

1. Activation directly by the rupture disc.
2. Activation by pressure or flow change.

Table 6-3 gives a brief description of the design principles of remote burst indicators.*

Figure 6-33 shows a simplified schematic of a rupture disc burst indicator which is designed on the flexible circuit principle. The heart of the indicator system is an alarm strip, which adheres to a PTFE membrane that is part of the rupture disc assembly. Upon bursting of the rupture disc, the alarm strip breaks and creates an open circuit which is detected by an alarm monitor. The monitor may be used, in addition, to activate or deactivate pumps, valves, or systems in response to the failure of the rupture disc.

Selection of Rupture Discs

The selection of rupture discs commences with a specification of the requirements. This specification may take the following form:

Operating conditions:

- Fluid
- Properties of fluid (molecular weight, ratio of specific heats, specific gravity, viscosity if high, tendency to produce deposits, etc.)
- Maximum allowable working pressure
- Normal operating pressure
- Operating temperature of disc
- Vacuum or back pressure
- Cycle conditions (magnitude and frequency of pressure excursions, temperature excursions)
- Required relieving capacity
- Hydraulic service (minimal or high flow rate after bursting)
- Other conditions

Application:

- Single disc installation, non-fire conditions
- Multiple disc installation, non-fire conditions

* Taken from "Rupture Disc Technical Manual" of Continental Disc Corporation, March 1, 1985.

Table 6-3
Design Principles of Remote Burst Indicators

Type	Description	Pros and Cons
Pressure Switch	Absolute pressure switch located downstream and senses the increase in pressure at time of rupture. Differential pressure switch across rupture disc (set for minimum pressure). Pressure switch located on pressure side of disc (Set just below burst pressure).	–Reasonably inexpensive. –Does not detect leaks. –Reusable. –Prone to clogging. –Usable on positive pressures only. –Requires pipe modification. –Special metals available. –Downstream switch cannot be used with header system. –Usable with all makes of rupture discs. –Will only work in closed systems.
Paddle Type Switch	A paddle switch is the paddle inserted in the downstream side of the rupture disc. As rupture occurs, the paddle is forced upward causing the switch to activate.	–Does not detect leaks. –Cannot be used on low pressures. –May be reusable if not damaged during rupture. Requires pipe modification. –Usable with all makes of rupture discs but is limited to small sized discs.
Excess Flow Switch	This switch is located on the downstream side of the disc. As rupture occurs, the pressure buildup pushes a ball into the switch thereby closing it.	–Does not detect leaks. –Cannot be used in a header system. –Reusable. –Subject to clogging. –Requires pipe modification. –Usable with all makes of rupture discs. –Will only work in closed systems.
Contact Opening Probe	Probe is secured to top of disc. As rupture occurs, probe is broken or pulled out of socket, thus opening a normally closed circuit.	–Can only be used with metallic discs. –Destroyed at rupture. –Special pipe/flange modification is needed.
Flexible Circuit	An electric circuit is attached to a disc component. The circuit is normally closed. Upon rupture, the circuit is broken and a signal is given.	–Can be used on C.D.C. rupture discs. –Can be used as leak detector in some cases. –Requires no pipe or flange modification. –Relatively inexpensive. –Fail-safe system

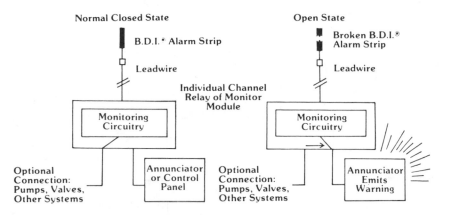

Figure 6-33. Simplified schematic of a rupture disc burst indicator designed on the flexible circuit principle. (Courtesy of Continental Disc Corporation.)

- Fire or external heat conditions
- Double disc assembly
- Rupture disc upstream of pressure relief valve
- Rupture disc downstream of pressure relief valve
- Rupture discs on both sides of pressure relief valve

Application code
Quality assurance documentation if required
Quantity required

With these data, the rupture disc may be preselected using a manufacturer's selection chart like the one shown in Table 6-4. The final selection should be made in cooperation with the rupture disc manufacturer.

With regard to the material specification for the disc, the user should attempt to give the manufacturer a choice of several types of material. This can be useful from the points of price and delivery time.

Reordering of Rupture Discs

When reordering rupture discs, the user should quote the lot or batch number to ensure the supply of the correct replacement discs. If the reordered discs are subject to a manufacturing range, quoting the stamped rating could lead to the supply of overrated discs. For example, if the original discs had been ordered for a bursting pressure of 10 bar (14.5 psi), and the agreed manufacturing range was plus/minus 10%, the disc could have been stamped/rated at 11 bar (16 psi). When reordering against the stamped rat-

Table 6-4
Rupture Disc Selection Chart

Disc Type	Burst Tolerance	MOP/RP	Manufacturing Range			Comment
			Burst Pressure psig.*	Under %	Over %	
* Standard Prebulged Solid Metal Rupture Disc	± 5%	70%	2 thru 8	40%	40%	70% MOP/RP at ambient temperature and static pressure.
			9 thru 12	30%	30%	
			13 thru 19	10%	20%	
			20 thru 50	4%	14%	
* Composite	± 5%	80%	51 thru 100	4%	10%	80% MOP/RP at ambient temperature and static pressure.
			101 thru 500	4%	7%	
			501 and up	3%	6%	
* Reverse Acting with Knife Blade	± 2% over 100 psig. ± 2 psig. below 100 psig	90%	15 thru 1000	None Required		
* Reverse Acting without Blades	± 5%	90%	20 thru 1000	0% 3 ranges 5% available 10%		MOP/RP varies with manufacturing range. One manufacturer has burst tolerance ± 5% over 60 psig & ± 3 psig under 60 psig
* Graphite Discs	± 5% over 15 psig. ± .75 psig from 1 to 14 psig.	70%	1 thru 450	None Required		
* Low Pressure Disc**	± 1/2" H2O on Max. specified Relief Press.	Within 1" H2O of Min. Rating	1 to 30" H2O	6" H2O Range		
* Vent Panel	± .5 psig	50%				Available in pressures from 1 to 4 psig.

Courtesy of Continental Disc Corporation.

* Dependent on Disc Size and Operating Temperature

** When low pressure disc is used as a composite disc reference composite disc characteristics.

ing, the replacement discs could be supplied with a stamped rating of, say, 12 bar (17.4 psi). The bursting pressure in this case could exceed the maximum allowable working pressure.

Installation and Maintenance

Table 6-5 gives guidelines for installation and maintenance of rupture discs. These guidelines must be followed to ensure satisfactory operation of the installation.

Table 6-5
Guidelines for Installation and Maintenance of Rupture Discs

		Do	Do Not
1.	Removal of shipping protection.	*Read instructions. *Inspect disc dome for damage *Inspect component parts for damage.	*Do not touch dome of disc. *Do not set disc dome side down.
2.	Carrying disc from one location to another.	*Handle disc by disc O.D. or flange or tag. *Transport disc in original packing if possible.	*Do not touch dome *Do not manhandle disc.
3.	Cleaning seal surfaces of disc and holder.	*Holder: Remove dirt, rust or product build-up from seat area, using cleaning solution as necessary. *Disc: Remove dirt or particles from seal surface flange area).	*Do not scrape or scratch holder sealing surface. *Do not clean dome of disc.
4.	Disc crown distortion.	*If disc has damaged dome, replace with new disc.	*Do not install damaged disc.
5.	Proper bolt torque.	*Read instructions for proper torque loading. *Use clean, lightly oiled studs and nuts.	*Do not use rusted, dirty or damaged studs or nuts.
6.	Installation of disc assembly.	*Install disc in direction shown by flow arrow shown on holder. *Use proper gasketing on on both sides of the holder (between holder and companion flange). *Check flow direction.	*Do not touch disc dome or hit dome against holder or other parts.

(table continued on next page)

Table 6-5
Continued

7. Installing knife blades (if required).	*Inspect blades for sharpness; if dull or damaged, replace.	*Do not use dull or damaged knife blades. *Do not install holder without knife blades, when blades are required. *Do not grind or machine blades or holder.
8. Checking for obstruction.	*Remove product buildup from blades. If sharpness of blades is affected, replace with new blades.	*Do not try to clean a disc that has product buildup. Replace with new disc.
9. Checking application.	*Check disc and flange pressure and temperature ratings. *Check disc for proper material and ratings.	*Do not install disc with a set pressure rating over the MAWP of the vessel. Consult factory.

Taken from "Rupture Disc Technical Manual" of Continental Disc Corporation, March 1, 1985.

User's Responsibility

The user must be fully aware of the installation and maintenance requirements of rupture discs. Under no circumstances should the installation be entrusted to a person who is unaware of the installation requirements and the consequences of improper handling of rupture discs.

Each plant should maintain a system which records location and specification of installed rupture discs, the dates and history of rupture disc failures, the agreed replacement periods, and ordering and storage instructions. The consequences of any misunderstanding can be enormous.

In general, only engineers trained in the selection, installation, and maintenance of rupture discs should be entrusted with specifying, procuring, and supervising the installation of rupture discs.

Standards Pertaining to Rupture Discs

A list of standards pertaining to rupture discs can be found in Appendix C.

Derivation of Flow Equations

Equation 5-20: Pressure Loss in Inlet Pipeline

The permissible pressure loss in the inlet pipeline is small only so that the change in volume of the gas can be neglected. The evaluation of the pressure loss may therefore be based on Darcy's equation for incompressible fluids:

$$\Delta P = \left(\Sigma \zeta + \frac{fl}{d_i} \right) \frac{v_i^2}{2V_i}$$

in which

$$v_i = \frac{WV_i}{A_i}$$

and, neglecting the change in volume of the gas:

$$V_i = V_1$$

But from the equation of state:

$$V_1 = \frac{RT_1}{P_1}$$

so that

$$\Delta P = \left(\Sigma \zeta + \frac{fl}{d_i} \right) \left(\frac{W}{A_i} \right)^2 \frac{RT_1}{2P_1}$$

Also from Equation 5-11:

$$W = K_d\,K_b\,A_n\,P_1 \left[\frac{1}{RT_1\,Z} \; k \left(\frac{2}{k+1} \right)^{(k+1)/(k-1)} \right]^{1/2} \tag{A-1}$$

$$\overset{*}{\underset{*}{*}}$$

$$\Delta P = \left(\Sigma\zeta + \frac{fl}{d_i} \right) \left(\frac{K_d K_b A_n}{A_i} \right)^2 \frac{P_1}{2Z} \; k \left(\frac{2}{k+1} \right)^{(k+1)/(k-1)} \tag{5-20}$$

Equation 5-21: Terminal Pressure P_t

From the continuity of mass flow:

$$v_t = \frac{WV_t}{A_t}$$

in which, from the equation of state:

$$V_t = \frac{RT_t}{P_t}$$

so that

$$P_t = \frac{WRT_t}{A_t\,v_t} \tag{A-2}$$

But for conditions of sonic outlet velocity

$$v_t = c_t = (kRT_t)^{1/2}$$

and isothermal flow

$$T_t = T_n$$

Equation A-1 may therefore also read

$$P_t = \frac{W}{A_t} \left(\frac{RT_n}{k} \right)^{1/2} \tag{A-3}$$

The value of T_n may be found from the constant energy equation for isentropic flow through the valve orifice

$$\frac{v_1^2}{2} + c_p\,T_1 = \frac{v_n^2}{2} + c_p\,T_n$$

The approach velocity for flow through a nozzle may be taken as zero, in which case

$$c_p \, T_1 \;=\; \frac{v_n^2}{2} + c_p \, T_n \tag{A-4}$$

But from the ideal gas laws

$$c_p \;=\; R \frac{k}{k-1}$$

so that Equation A-3 may also read

$$\frac{k}{k-1} RT_1 \;=\; \frac{v_n^2}{2} + \frac{k}{k-1} RT_n$$

or, rearranged:

$$v_n^2 \;=\; 2R \frac{k}{k-1} \, (T_1 - T_n)$$

and

$$T_1 \;=\; T_n + \frac{(k-1)\, v_n^2}{2kR} = T_n \left[1 + \left(\frac{k-1}{2} \right) \frac{v_n^2}{kRT_n} \right]$$

But

$$kRT_n = c_n^2 \quad \text{and} \quad \frac{v_n}{c_n} = N_n$$

so that

$$T_1 \;=\; T_n \left(1 + \frac{k-1}{2} N_n^2 \right) \tag{A-5}$$

and

$$T_n \;=\; \frac{T_1}{1 + \dfrac{k-1}{2} N_n^2} \tag{A-6}$$

But for
$$N_n = 1.0$$

$$T_n \;=\; \frac{T_1}{1 + \dfrac{k-1}{2}} = T_1 \frac{2}{k+1} \tag{A-7}$$

Introducing this expression in Equation A-3:

$$P_t \;=\; \frac{W}{A_t} \left(\frac{2RT_1}{k\,(k+1)} \right)^{1/2}$$

Substituting W with the expression of Equation A-1:

$$P_t = \frac{K_d K_b A_n P_1}{A_t} \left(\frac{2}{k+1}\right)^{k/(k-1)} \left(\frac{1}{Z}\right)^{1/2}$$

(5-21)

If the pressure thus calculated is lower than P_3, the flow velocity at the outlet is subsonic, in which case

$$P_t \quad = P_3$$

Equation 5-22: Back Pressure P_2

From the law of conservation of energy:

$$\frac{2V}{v^2} dP + \frac{2dv}{v} + \Sigma\zeta + \frac{fl}{d} = 0$$

The expression for $\dfrac{V}{V^2}$ in the above equation may been found from the equation of state for conditions of constant temperature:

$$PV \quad = P_t V_t.$$

and from the relationship for continuity of mass flow:

$$\frac{v}{v_t} = \frac{V}{V_t} \text{ or } v^2 V_t^2 = v_t^2 V^2$$

Thus

$$\frac{P}{v_t^2 V} = \frac{P_t}{v^2 V_t}$$

and transformed

$$\frac{V}{v^2} = \frac{PV_t}{P_t v_t^2}$$

The constant energy equation may therefore be rewritten:

$$\frac{2V_t}{P_t v_t^2} PdP + \frac{2dv}{v} + \Sigma\zeta + \frac{fl}{d} = 0$$

and integrated between the limits of 2 and t:

$$\frac{2V_t}{P_t v_t^2} \left(\frac{P_t^2 - P_2^2}{2}\right) + 2 \ln \frac{v_t}{v_2} + \Sigma\zeta + \frac{fl}{d} = 0$$

so that

$$P_2 = \left[P_t^2 + \left(2 \ln \frac{v_t}{v^2} + \Sigma \zeta + \frac{fl}{d} \right) \frac{P_t v_t^2}{V_t} \right]^{1/2}$$

where $d = d_t$.

But

$$\frac{v_t^2}{V_t} = \left(\frac{W}{A_t} \right)^2 V_t$$

in which, from the equation of state:

$$V_t = \frac{RT_t}{P_t}$$

and from Equation A-7 under consideration of isothermal pipeline flow

$$T_t = T_1 \frac{2}{k+1}$$

so that

$$\frac{v_t^2}{V_t} = \left(\frac{W}{A_t} \right)^2 \frac{RT_1}{P_:} \frac{2}{k+1}$$

Also, from the continuity of flow

$$v_2 = v_t \frac{V_2}{V_t}$$

in which, from the equation of state for constant temperature:

$$\frac{V_2}{V_t} = \frac{P_t}{P_2}$$

so that

$$2 \ln \frac{v_t}{v_2} = 2 \ln \frac{P_2}{P_t}$$

and

$$P_2 = \left[P_t^2 + \left(2 \ln \frac{P_2}{P_t} + \Sigma \zeta + \frac{fl}{d_t} \right) \left(\frac{W}{A_t} \right)^2 RT_1 \frac{2}{k+1} \right]^{1/2}$$

Substituting W with the expression of Equation A-1:

$$P_2 = \left[P_t^2 + \left(2 \ln \frac{P_2}{P_t} + \Sigma \zeta + \frac{fl}{d_t} \right) \right.$$

$$\left. \left(\frac{K_d K_b A_n P_1}{A_t} \right)^2 \frac{k}{Z} \left(\frac{2}{k+1} \right)^{2k/(k-1)} \right]^{1/2} \tag{5-22}$$

Equation 5-23: Diffuser Stack Area A_t for Mach Number $N_t < 1.0$

From Equation A-2:
$$A_t = \frac{W R T_t}{P_t v_t}$$

in which, for $N_t < 1.0$:
$$P_t = P_3$$

Also
$$v_t = c_t N_t = (k R T_t)^{1/2} N_t$$

and for isothermal pipeline flow
$$T_t = T_{sn}$$

so that
$$A_t = \frac{W}{P_3 N_t} \left(\frac{R T_{sn}}{k} \right)^{1/2} \tag{A-8}$$

But analogous to Equation A-6
$$T_{sn} = \frac{T_4}{1 + \dfrac{k-1}{2} N_{sn}^2}$$

and from Equation A-7 for isothermal pipeline flow
$$T_n = T_4 = T_1 \frac{2}{k+1}$$

so that
$$T_{sn} = \frac{2 T_1}{(k+1) \left(1 + \dfrac{k-1}{2} N_{sn}^2 \right)} \tag{A-9}$$

Substituting above in Equation A-8:

$$A_t = \frac{W}{P_3 N_t} \left(\frac{2RT_1}{k(k+1)(1+\frac{k-1}{2}N_{sn}^2)} \right)^{1/2}$$

Or, substituting W with the expression of equation A-1:

$$A_t = \frac{K_d K_b A_n P_1}{P_3 N_t} \left(\frac{2}{k+1} \right)^{k/(k-1)} \left(\frac{1}{Z(1+\frac{k-1}{Z}N_{sn}^2)} \right)^{1/2} \qquad (5\text{-}23)$$

Equation 5-24: Back Pressure P_s at Outlet of Diffuser Nozzles

The pressure P_s consists of the pressure P_3 and the pressure loss in the outlet stack. In calculating this pressure loss, the specific volume of the gas flowing through the stack may be assumed to be constant because of the low flow velocities involved. The evaluation of the pressure loss may be based therefore on Darcy's equation for incompressible fluids, so that:

$$P_s = P_3 + \Delta P = P_3 + \left(\Sigma \zeta + \frac{fl}{d_t} \right) \frac{v_t^2}{2V_t}$$

in which

$$v_t = \frac{WV_t}{A_t}$$

and from the equation of state:

$$V_t = \frac{RT_t}{P_t}$$

But for subsonic flow velocity at the outlet of the stack:
$$P_t = P_3$$

and for conditions of isothermal flow in the stack:
$$T_t = T_{sn}$$

T_{sn} may be found from Equation A-9.

$$** P_s = P_3 + \left(\Sigma \zeta + \frac{fl}{d_t} \right) \left(\frac{W}{A_t} \right)^2 \frac{RT_1}{P_3(k+1)(1+\frac{k-1}{2}N_{sn}^2)}$$

Substituting W with the expression of Equation A-1:

$$P_s = P_3 + \left(\Sigma\zeta + \frac{fl}{d_t}\right)\left(\frac{K_d K_b A_n P_1}{A_t}\right)^2$$

$$\frac{k}{2P_3 Z(1 + \dfrac{k-1}{2} N_{sn}{}^2)}\left(\frac{2}{k+1}\right)^{2k/(k-1)} \tag{5-24}$$

Equation 5-25: Diffuser Nozzle Area A_{sn} for Mach Number $N_{sn} < 1.0$

From Equation A-2, but for flow section "sn"

$$A_{sn} = \frac{WRT_{sn}}{P_{sn}\, v_{sn}}$$

in which

$$v_{sn} = c_{sn}\, N_{sn} = (kRT_{sn})^{1/2}\, N_{sn}$$

and for Mach number $N_{sn} < 1.0$:

$$P_{sn} = P_s$$

then

$$A_{sn} = \frac{W}{P_s\, N_{sn}}\left(\frac{RT_{sn}}{k}\right)^{1/2}$$

Substituting in above the expression for T_{sn} from Equation A-9, and introducing the coefficient of discharge K_z for flow through the diffuser nozzles:

$$A_{sn} = \frac{W}{N_{sn}\, P_s\, K_z}\left(\frac{2RT_1}{k\,(k+1)\,(1 + \dfrac{k-1}{2}\, N_{sn}{}^2)}\right)^{1/2}$$

Substituting W with the expression of Equation A-1:

$$A_{sn} = \frac{K_d\, K_b\, A_n\, P_1}{K_z\, N_{sn}\, P_s}\left(\frac{2}{k+1}\right)^{k/(k-1)}\left(\frac{1}{Z\,(1 + \dfrac{k-1}{2} N_{sn}{}^2)}\right)^{1/2} \tag{5-25}$$

Equation 5-26: Pressure P_4 at Inlet of Diffuser Nozzles

The pressure P_4 may be found from the expression for the Mach number

$$N_{sn} = \frac{v_{sn}}{c_{sn}}$$

in which for isentropic nozzle flow*

$$v_{sn} = \left\{ \frac{2k}{k-1} P_4 V_4 \left[1 - \left(\frac{P_s}{P_4} \right)^{(k-1)/k} \right] \right\}^{1/2}$$

and*

$$c_{sn} = \left(\frac{2k}{k+1} P_4 V_4 \right)^{1/2}$$

so that

$$N_{sn} = \left\{ \frac{k+1}{k-1} \left[1 \left(\frac{P_s}{P_4} \right)^{(k-1)/k} \right] \right\}^{1/2}$$

and, rearranged:

$$P_4 = \frac{P_s}{(1 - \frac{k-1}{k+1} N_{sn}^2)^{\,k/(k-1)}} \tag{5-26}$$

Equation 5-30: Discharge Reactive Force for Subsonic Discharge Velocity

The equation for the discharge reactive force is derived from the momentum equation:

$$- A_t (P_t - P_3) + F = Wv_t$$

Therefore the general equation for the discharge reactive force is

$$F = Wv_t + A_t (P_t - P_3)$$

in which, for gases or vapors at subsonic discharge velocity:

$$P_t = P_3$$

But from the continuity of mass flow

$$v_t = \frac{WV_t}{A_t}$$

*Refer to any textbook on thermodynamics.

and from the equation of state:

$$V_t = \frac{RT_t}{P_t} = \frac{RT_t}{P_3}$$

Also for isothermal flow

$$T_t = T_n$$

and from Equation A-7

$$T_n = T_1 \frac{2}{k+1}$$

*
**

$$F = \frac{W^2 RT_1}{A_t P_3} \left(\frac{2}{k+1} \right)$$

or, substituting W with the expression of Equation A-1:

$$F = \frac{K_d K_b A_n P_1}{A_t P_3 Z} k \left(\frac{2}{k+1} \right)^{2k/(k+1)} \tag{5-30}$$

Equation 5-31: Discharge Reactive Force for Sonic Discharge Velocity

The equation for the discharge reactive force for gases or vapors discharging at sonic velocity may be written:

$$F = Wc_t + A (P_t - P_3)$$

in which

$$c_t = (kRT_t)^{1/2}$$

and

$$T_t = T_1 \frac{2}{k+1}$$

so that

$$F = W \left(\frac{2k}{k+1} RT_1 \right)^{1/2} + A (P_t - P_3)$$

Substituting W with the expression of Equation A-1, and P_t with the expression of Equation 5-21 and combining, the equation for the discharge reactive force for sonic discharge velocity may be written:

$$F = (1+k) \left[K_d K_b A_n P_1 \left(\frac{2}{k+1} \right)^{k/(k-1)} \left(\frac{1}{Z} \right)^{1/2} \right] - A_t P_3 \tag{5-31}$$

Properties of Fluids

This section contains the following:

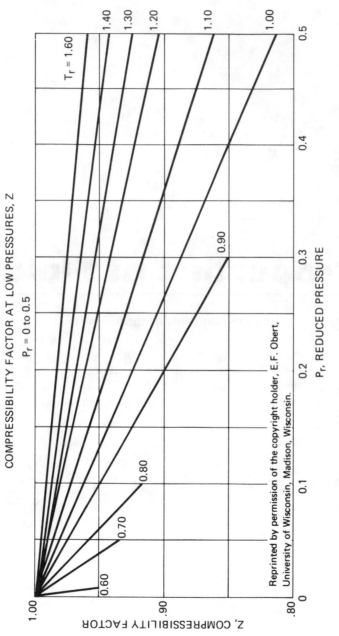

COMPRESSIBILITY FACTOR AT LOW PRESSURES, Z

P_r = 0 to 0.5

T_r = 1.60

1.40
1.30
1.20
1.10
1.00

0.90

0.80

0.70

0.60

Z, COMPRESSIBILITY FACTOR

P_r, REDUCED PRESSURE

Reprinted by permission of the copyright holder, E.F. Obert,
University of Wisconsin, Madison, Wisconsin.

Figure B-1. Compressibility factor, Z, at low pressures for P_r = 0 to 0.5.

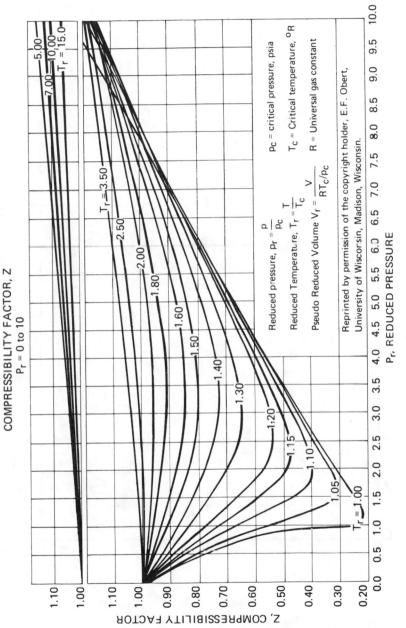

Figure B-2. Compressibility factor, Z, for P_r = 0 to 10.

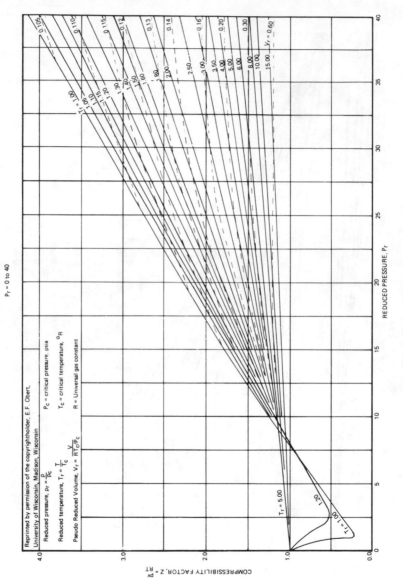

$P_r = 0$ to 40

Reprinted by permission of the copyrightholder, E. F. Obert,
University of Wisconsin, Madison, Wisconsin

P_c = critical pressure, psia

T_c = critical temperature, °R

R = Universal gas constant

Reduced pressure, $p_r = \dfrac{p}{p_c}$

Reduced temperature, $T_r = \dfrac{T}{T_c}$

Pseudo Reduced Volume, $V_r = \dfrac{V}{RT_c/p_c}$

REDUCED PRESSURE, P_r

COMPRESSIBILITY FACTOR, $Z = \dfrac{pv}{RT}$

Figure B-3. Compressibility factor, Z, for P_r = 0 to 40.

Table B-1
Physical Constants of Gases

Gas		M Molecular Weight	R Individual Gas Constant		Isentropic Coefficient k for P – O T = 273K = 492R	T_c Critical Temperature		P_c Critical Pressure (absolute)	
			J kg.K	ft. lbf lb.°R		°K	°R	MPa	lb/in²
Acetylene	C_2H_2	26.078	318.82	59.24	1.23	309.09	556.4	6.237	904.4
Air	–	28.96	287.09	53.35	1.40	132.4	238.3	3.776	547.5
Ammonia	NH_3	17.032	488.15	90.71	1.31	405.6	730.1	11.298	1638.2
Argon	Ar	39.994	208.15	38.68	1.55	150.8	271.4	4.864	705.3
Benzol	C_6H_6	78.108	106.44	19.78	–	561.8	1011.2	4.854	703.9
Butane-n	C_4H_{10}	58.124	143.04	26.58	–	425.2	765.4	3.506	508.4
Butan-i	C_4H_{10}	58.124	143.04	26.58	–	408.13	734.6	3.648	529.0
Butylene	C_4H_8	56.108	148.18	27.54	–	419.55	755.2	3.926	569.2
Carbon dioxide	CO_2	44.011	188.91	35.10	1.30	304.2	547.6	7.385	1070.8
Carbon disulfide	CS_2	76.142	109.19	20.29	–	546.3	983.3	7.375	1069.4
Carbon monoxide	CO	28.011	296.82	55.16	1.40	133.0	239.4	3.491	506.2

(continued)

Table B-1
Continued

Gas		M Molecular Weight	R Individual Gas Constant		Isentropic Coefficient k for $P \to O$ $T = 273K$ $= 492R$	T_c Critical Temperature		P_c Critical Pressure (absolute)	
			J kg.K	ft. lbf lb.°R		°K	°R	MPa	lb/in²
Carbon oxysulfide	COS	60.077	138.39	25.72	-	375.35	675.6	6.178	895.9
Chlorine	Cl_2	70.914	117.24	21.79	1.34	417.2	751.0	7.698	1116.3
Cyanogen	C_2N_2	52.038	159.77	29.69	-	399.7	719.5	5.894	854.6
Ethane	C_2H_6	30.070	276.49	51.38	1.20	305.42	549.8	4.884	708.2
Ethylene	C_2H_4	28.054	296.36	55.07	1.25	282.4	508.3	5.070	735.2
Helium	He	4.003	2076.96	385.95	1.63	5.2	9.4	0.229	33.22
Hydrogen	H_2	2.016	4124.11	766.36	1.41	33.3	59.9	1.295	187.7
Hydrogen chloride	HCl	36.465	228.01	42.37	1.39	324.7	584.5	8.307	1204.4
Hydrogen cyonide	HCN	27.027	307.63	57.16	-	456.7	822.1	5.394	782.1
Hydrogen sulfide	H_2S	34.082	243.94	45.33	1.33	373.53	672.4	9.013	1306.8

(continued)

Gas		M Molecular Weight	R Individual Gas Constant		Isentropic Coefficient k for P → 0 T = 273K = 492R	T_c Critical Temperature		P_c Critical Pressure (absolute)	
			J kg.K	ft.lbf lb.°R		°K	°R	MPa	lb/in²
Methane	CH_4	16.034	518.24	96.30	1.31	190.7	343.3	4.629	671.2
Methyl chloride	CH_3Cl	50.491	164.66	30.60	–	416.2	749.2	6.669	667.0
Neon	Ne	20.183	411.94	76.55	1.64	44.4	79.9	2.654	384.8
Nitric oxide	NO	30.008	277.06	51.48	1.39	180.2	324.4	6.541	948.5
Nitrogen	N_2	28.016	296.76	55.15	1.40	126.3	227.3	3.383	490.6
Nitrous oxide	N_2O	44.016	188.89	35.10	1.28	309.7	557.5	7.267	1053.7
Oxygen	O_2	32.000	259.82	48.28	1.40	154.77	278.6	5.080	736.6
Sulfur dioxide	SO_2	64.066	125.77	24.11	1.28	430.7	775.3	7.885	1143.3
Propane	C_3H_8	44.097	188.54	35.04	–	370.0	666.0	4.256	617.15
Propylene	C_3H_6	42.081	197.56	36.71	–	364.91	656.8	4.621	670.1
Tuluene	C_7H_8	92.134	90.24	16.77	–	593.8	1068.8	4.207	610.0
Water vapour	H_2O	18.016	461.48	85.75	1.33 *	647.3	1165.1	22.129	3208.8
Xylene	C_8H_{10}	106.16	78.32	14.55	–	–	–	–	–

* at 100°C

Source: VDI Blatt 4, Entwurf, Jan. 1970, Berechnungsgrundlagen für die Durchflussmessung mit Drosselgeräten, Stoffwerte.

By Courtesy of VDI/VDE – Gesellschaft für Mess–und Regelungstechnik.

Appendix
C

Standards
Pertaining to Valves

This chapter lists common USA and British standards pertaining to valves, as published in the standard indexes of the various standard organizations for 1980. Because new standards are continually issued and old standards revised or withdrawn, the validity of these standards should be verified prior to application.

Standard Organizations

ANSI   American National Standards Institute
1430 Broadway
New York, N.Y. 10018

API   American Petroleum Institute
2101 L Street, N.W.
Washington, D.C. 20037

ASME   The American Society of Mechanical
Engineers
United Engineering Center
345 East 47th Street
New York, N.Y. 10017

AWWA   American Water Works Association
6666 West Quincy Avenue
Denver, Colorado 80235

MSS   Manufacturers Standardization Society of the
                                Valves and Fittings Industry, Inc.
                                5203 Leesburg Pike, Suite 502
                                Falls Church, Virginia 22041

BSI   British Standards Institution
                                2 Park Street
                                London W1A 2BS

Standards Pertaining to Valve Ends

MSS SP-6   Standard finishes for contact faces of pipe
                                flanges and connecting-end flanges of valves
                                and fittings.

MSS SP-9   Spot facing for bronze, iron, and steel flanges.

MSS SP-44   Steel pipeline flanges.

MSS SP-51   150 lb corrosion-resistant cast flanges and
                                flanged fittings.

MSS SP-65   High-pressure chemical industry flanges and
                                threaded stubs for use with lens gaskets.

MSS-84   Steel valves — socket welding and threaded
                                ends.

API Spec 6A   Specification for wellhead equipment (includ-
                                ing flanges).

API Std 605   Large diameter carbon steel flanges.

ANSI B1.20.3   Dryseal pipe threads (inch).

ANSI B2.1   Pipe threads (except dryseal).

ANSI B16.1   Cast iron pipe flanges and flanged fittings, 25,
                                125, 250, and 800 lb.

ANSI B16.5   Steel pipe flanges and flanged fittings, includ-
                                ing ratings for class 150, 300, 400, 600, 900,
                                1500, and 2500.

ANSI B16.20   Ring joint gaskets and grooves for steel pipe
                                flanges.

ANSI B16.24   Bronze flanges and flanged fittings, 150 and
                                300 lb.

ANSI B16.25   Butt-welding ends.

ANSI B16.31   Non-ferrous pipe flanges, 150, 300, 400, 600,
                                900, 1500, and 2500 lb.

ANSI/AWWA C207-78 — Flanges for water-works service, 4 in through 144 in, steel.

ANSI/AWWA C606-78 — Joints, grooved and shouldered type.

BS 10 — Flanges and bolting for pipes, valves, and fittings (obsolescent).

BS 21 — Pipe threads for tubes and fittings where pressure-tight joints are made on the threads.

BS 1560 — Steel pipe flanges and flanged fittings (nominal sizes ½ in to 24 in) for the petroleum industry.
Part 2 (1970), metric dimensions.

BS 3293 — Carbon steel flanges (over 24 in nominal size) for the petroleum industry.

BS 4504 — Flanges and bolting for pipes, valves, and fittings, metric series.
Part 1 (1969), ferrous.
Part 2 (1974), copper alloy and composite flanges.

Standards Pertaining to Globe Valves

MSS SP-42 — Corrosion-resistant gate, globe, angle, and check valves with flanged and butt-weld ends.

MSS SP-61 — Pressure testing of steel valves.

MSS SP-80 — Bronze gate, globe, angle, and check valves.

MSS SP-84 — Steel valves — socket welding and threaded ends.

MSS SP-85 — Cast iron globe and angle valves, flanged and threaded ends.

API RP 6F — Recommended practice for fire test for valves (tentative).

ANSI B16-10 — Face-to-face and end-to-end dimensions of ferrous valves.

ANSI B16.34 — Steel valves, flanged and butt-welding end.

BS 1873 — Steel globe valves and stop and check valves (flanged and butt-welding ends), for the petroleum, petrochemical, and allied industries.

BS 2080	Face-to-face, center-to-face, end-to-end, and center-to-end dimensions of flanged and butt-welding end steel valves, for the petroleum, petrochemical, and allied industries.
BS 2996	Cast and forged steel wedge gate, globe, check, and plug valves, screwed and socket welding, sizes 2 in and smaller, for the petroleum industry.
BS 5146	Inspection and test of steel valves for the petroleum, petrochemical, and allied industries.
BS 5152	Cast iron globe and globe stop and check valves, for general purposes.
BS 5154	Copper alloy globe, globe stop and check, check, and gate valves (including parallel slide type), for general purposes.
BS 5160	Specification for flanged steel globe valves, globe stop and check valves, and lift-type check valves for general purposes.
BS 5417	Testing of general purpose and industrial valves.
BS MA 65 (for marine pipeline systems)	
Part 5	Cast iron globe and globe stop and check valves.
Part 6	Steel globe stop and check valves.
Part 9	Copper alloy globe, globe stop and check, check, and gate valves.

Standards Pertaining to Piston Valves

Standards of the organizations listed in earlier in this appendix, page do not deal specifically with piston valves.

Standards Pertaining to Parallel and Wedge Gate Valves

| MSS SP-42 | Corrosion-resistant gate, globe, angle, and check valves with flanged and butt-weld ends. |
| MSS SP-45 | Bypass and drain connection standard. |

MSS SP-61	Hydrostatic testing of steel valves.
MSS SP-70	Cast iron gate valves, flanged and threaded ends.
MSS SP-80	Bronze gate, globe, angle, and check valves.
MSS SP-81	Stainless steel, bonnetless, flanged, wafer, knife gate valves.
API Spec 6D	Specification for pipeline valves, end closures, connectors, and swivels.
API RP 6F	Recommended practice for fire test for valves (tentative).
API Std 595	Cast iron gate valves, flanged ends.
API Std 597	Steel venturi gate valves, flanged or butt-welding ends.
API Std 598	Valve inspection and test.
API Std 600	Steel gate valves, flanged and butt-welding ends.
API Std 602	Compact carbon steel gate valves.
API Std 603	Class 150, corrosion-resistant gate valves.
API Std 604	Ductile iron gate valves, flanged ends.
API Std 605	Compact carbon steel gate valves (extended body).
ANSI B16.10	Face-to-face and end-to-end dimensions of ferrous valves.
ANSI B16.34	Steel valves, flanged and butt-welding end.
BS 1414	Steel wedge gate valves (flanged and butt-welding ends) for the petroleum, petrochemical and allied industries.
BS 1952	Copper alloy gate valves for general purposes.
BS 2080	Face-to-face, center-to-face, end-to-end, and center-to-end dimensions of flanged and butt-welding end steel valves for the petroleum, petrochemical and allied industries.
BS 2995	Cast and forged steel wedge gate, globe, check, and plug valves, screwed and socket welding, sizes 2 in and smaller, for the petroleum industry.

BS 5146	Inspection and test of steel valves for the petroleum, petrochemical and allied industries.
BS 5150	Cast iron wedge and double-disc gate valves, for general purposes.
BS 5151	Cast iron parallel slide gate valves, for general purposes.
BS 5154	Copper alloy globe, globe stop and check, check, and gate valves (including parallel slide type), for general purposes.
BS 5157	Steel parallel slide gate valves, for general purposes.
BS 5163	Double-flanged cast iron wedge gate valves for water works purposes.
BS 5417	Testing of general purpose industrial valves.
BS MA 65 (for marine pipeline systems)	
Part 1	Cast iron gate valves.
Part 2	Steel gate valves.
Part 3	Cast iron parallel slide valves.
Part 4	Steel parallel slide valves.
BS MA 68	Gate valves for cargo oil systems.

Standards Pertaining to Plug Valves

MSS SP-61	Pressure testing of steel valves.
MSS SP-78	Cast iron plug valves, flanged and threaded ends.
MSS SP-84	Steel valves — socket welding and threaded ends.
API Spec 6A	Specification for wellhead equipment.
API Spec 6D	Specification for pipeline valves, end closures, connectors, and swivels.
API RP 6F	Recommended practice for fire test for valves (tentative).
API Std 593	Ductile iron plug valves, flanged ends.
API Std 599	Steel plug valves, flanged and butt-welding ends.

ANSI B16.10	Face-to-face and end-to-end dimensions of ferrous valves.
ANSI B16.34	Steel valves, flanged and butt-welding end.
BS 2080	Face-to-face, center-to-face, end-to-end, and center-to-end dimensions of flanged and butt-welding end steel valves for the petroleum, petrochemical, and allied industries.
BS 5146	Inspection and test of steel valves for the petroleum, petrochemical, and allied industries.
BS 5158	Cast iron and cast steel plug valves for general purposes.
BS 5353	Specification for plug valves.
BS 5417	Testing of general purpose industrial valves.

Standards Pertaining to Ball Valves

MSS SP-61	Pressure testing of steel valves.
MSS SP-72	Ball valves with flanged or butt-welding ends for general service.
MSS SP-84	Steel valves, socket welding and threaded ends.
API Spec 6D	Specification for pipeline valves, end closures, connectors, and swivels.
API Std 598	Valve inspection and test.
API Std 607	Fire test for soft-seated ball valves (tentative).
ANSI B16.10	Face-to-face and end-to-end dimensions of ferrous valves.
ANSI B16.34	Steel valves, flanged and butt-welding end.
BS 2080	Face-to-face, center-to-face, end-to-end, and center-to-end dimensions of flanged and butt-welding end steel valves for the petroleum, petrochemical, and allied industries.
BS 5159	Cast iron carbon steel ball valves for general purposes.
BS 5351	Steel ball valves for the petroleum, petrochemical, and allied industries.
BS 5417	Testing of general purpose industrial valves.

Standards Pertaining to Butterfly Valves

MSS SP-67	Butterfly valves.
API Std 598	Valve inspection and test.
API Std 609	Butterfly valves, lug-type and wafer-type.
ANSI/AWWA C504-80	Rubber-seated butterfly valves.
BS 3952	Cast iron butterfly valves for general purposes.
BS 5155	Cast iron and carbon steel butterfly valves for general purposes.
BS 5417	Testing of general purpose industrial valves.
BS MA65 (for marine pipeline systems) Part 10	Butterfly valves.

Standards Pertaining to Pinch Valves

Standards of the organizations listed earlier in the appendix, page 286, do not deal specifically with pinch valves.

Standards Pertaining to Diaphragm Valves

MSS SP-88	Diaphragm-type valves.
MSS SP-25	Standard marking system for valves, fittings, flanges and unions.
BS 5156	Screwdown diaphragm valves for general purposes.
BS 5417	Testing of general purpose industrial valves.
BS 5418	Marking of general purpose industrial valves.
ISO 5209	Marking of general purpose industrial valves.
DIN 3359	*Membran-Absperrarmaturen aus metallischen Werkstoffen.*

Standards Pertaining to Stainless Steel Valves

MSS SP-42	Corrosion-resistant gate, globe, angle, and check valves with flanged and butt-weld ends.
API Std 603	Class 150, corrosion-resistant gate valves.

Standards Pertaining to Check Valves

MSS SP-42	Corrosion-resistant gate, globe, angle, and check valves with flanged and butt-weld ends.
MSS SP-61	Pressure testing of steel valves.
MSS SP-71	Cast iron swing check valves, flanged and threaded ends.
MSS SP-80	Bronze gate, globe, angle, and check valves.
MSS SP-84	Steel valves — socket welding and threaded ends.
API Spec 6D	Specification for pipeline valves, end closures, connectors, and swivels.
API RP 6F	Recommended practice for fire test for valves (tentative).
API Std 594	Wafer-type check valves.
ANSI B16.10	Face-to-face and end-to-end dimensions of ferrous valves.
ANSI B16.34	Steel valves, flanged and butt-welding end.
BS 1868	Steel check valves (flanged and butt-welding ends) for the petroleum, petrochemical, and allied industries.
BS 1873	Steel globe and globe stop and check valves (flanged and butt-welding ends) for the petroleum, petrochemical, and allied industries.
BS 1953	Copper alloy check valves for general purposes.
BS 2080	Face-to-face, center-to-face, end-to-end, and center-to-end dimensions for flanged and butt-welding end steel valves for the petroleum, petrochemical, and allied industries.
BS 2995	Cast and forged steel wedge gate, globe, check, and plug valves, screwed and socket-welding, sizes 2 in and smaller, for the petroleum industry.
BS 4090	Cast iron check valves for general purposes.
BS 5146	Inspection and test of steel valves for the petroleum, petrochemical, and allied industries.
BS 5152	Cast iron globe and globe stop and check valves for general purposes.

BS 5153	Cast iron check valves for general purposes.
BS 5154	Copper alloy globe, globe stop and check, check, and gate valves (including parallel slide type) for general purposes.
BS 5160	Specification for flanged steel globe valves, globe stop and check valves, and lift-type check valves for general purposes.
BS 5417	Testing of general purpose industrial valves.
MS MA 65 (for marine pipeline systems)	
Part 5	Cast iron globe and globe stop and check valves.
Part 6	Steel globe stop and check valves.
Part 7	Cast iron check valves.
Part 8	Steel check valves.
Part 9	Copper alloy globe, globe stop and check, check, and gate valves.

Standards Pertaining to Pressure Relief Valves

API RP 520	Recommended practice for the design and installation of pressure relieving systems in refineries. Part I (1976) — Design Part II (1973) — Installation.
API RP 521	Guide for pressure relief and depressurizing systems.
ANSI/API 526	Flanged-steel safety relief valves.
ANSI/API 527	Commercial seat tightness of safety relief valves with metal-to-metal seats.

Standards for the Inspection and Testing of Valves

MSS SP-61	Pressure testing of steel valves.
MSS SP-82	Valve-pressure testing methods.
API RP 6F	Recommended practice for fire test for valves (tentative).
ANSI/API 527	Commercial seat tightness of safety relief valves with metal-to-metal seats.

API Std 598	Valve inspection and test.
API Std 607	Fire test for soft-seated ball valves (tentative).
BS 3636	Methods for proving the gas tightness of vacuum or pressurized plant.
BS 5146	Inspection and test of steel valves for the petroleum, petrochemical, and allied industries.

Miscellaneous Standards Pertaining to Valves

| BS 4371 | Fibrous gland packings. |

Standards Pertaining to Rupture Discs

ASME Code, Section VIII, Division 1, UG 125 through 136	
BS 2915	Bursting discs and bursting-disc devices.
ISO 6718	Bursting discs and bursting-disc devices.
ANSI/NFPA 68	Explosion venting.
VDI 3673	Pressure release of dust explosions.

Appendix D

International System of Units (SI)

SI Units

The international system of units is based upon:

1. Seven base units (Table D-1);
2. Two supplementary units (Table D-2);
3. Derived units.

The derived units may be divided into three groups:

1. Units which are expressed in terms of base and supplementary units (Table D-3);
2. Units which have been given special names and symbols (Table D-4);
3. Units which are expressed in terms of other derived units (Table D-5).

Decimal multiples and sub-multiples may be formed by adding prefixes to the SI units (Table D-6).

SI Units Conversion Factors

Table D-7 gives the conversion factors for Imperial, metric, and SI units.

Table D-1
Base Units of SI

length	meter	m
mass	kilogram	kg
time	second	s
electric current	ampere	A
temperature	kelvin	K
luminous intensity	candela	cd
amount of substance	mole	mol

Table D-2
Supplementary Units of SI

plane angle	radian	rad
solid angle	steradian	sr

Table D-3
Some Derived Units Expressed in Terms of Base and Supplementary Units

acceleration	meter per second squared	m/s^2
angular acceleration	radian per second squared	rad/s^2
area	square meter	m^2
coefficient of linear expansion	1 per kelvin	$1/K$
density	kilogram per cubic meter	kg/m^3
kinematic viscosity	square meter per second	m^2/s
mass flow rate	kilogram per second	kg/s
molar mass	kilogram per mole	kg/mol
specific volume	cubic meter per kilogram	m^3/kg
velocity	meter per second	m/s
volume	cubic meter	m^3

Table D-4
Some Derived Units Having Special Names

force	newton N	1 N = 1 kg·m/s²	
pressure stress	pascal Pa	1 Pa = 1 N/m²	= 1 kg/m·s²
energy work quantity of heat radiant energy	joule J	1 J = 1 N·m	= 1 kg·m²/s²
power radiant flux	watt W	1W = 1 J/s	= 1 kg·m²/s³
potential difference electromotive force electric potential	volt V	1 V = 1 W/A	= 1 kg·m²/A·s³

Table D-5
Some Derived Units Expressed in Terms of Other Derived Units

dynamic viscosity	Pa·s	=	kg/m·s
entropy	J/K	=	kg·m²/s²·K
heat capacity	J/K	=	kg·m²/s²·K
heat flux density	W/m²	=	kg/s³
molar energy	J/mol	=	kg·m²/s²·mol
molar entropy	J/mol·K	=	kg·m²/s²·K mol
molar heat capacity	J/mol·K	=	kg·m²/s²·K mol
moment of force	N·m	=	kg·m²/s²
radiant intensity	W/sr	=	kgm²/s³sr
specific energy	J/kg	=	m²/s²
specific entropy	J/kg·K	=	m²s²·K
specific heat capacity	J/kg·K	=	m²/s²·K
specific latent heat	J/kg	=	m²/s²
surface tension	N/m	=	kg/s²
torque	N·m	=	kg·m²/s²

Table D-6
Some SI Prefixes

10^9	giga	G
10^6	mega	M
10^3	kilo	k
10^2	hecto	h
10	deka	da
10^{-1}	deci	d
10^{-2}	centi	c
10^{-3}	milli	m
10^{-6}	micro	μ
10^{-9}	nano	n

Table D-7
Imperial, Metric, and SI Units Conversion Factors
Length

mm	cm	in	ft	yd	m	km	mile
1*	0.1*	0.0393701	3.2808×10^{-3}	1.0936×10^{-3}	10^{-3}*		
10*	1*	0.393701	0.032808	0.010936	0.01*		
25.4*	2.54*	1*	0.083333	0.027778	0.0254*		
304.8*	30.48*	12*	1*	0.333333	0.3048*	3.048×10^{-4}*	1.894×10^{-4}
914.4*	91.44*	36*	3*	1*	0.9144*	9.144×10^{-4}*	5.682×10^{-4}
1000*	100*	39.3701	3.28084	1.09361	1*	10^{-3}*	6.214×10^{-4}
10^6*	100000*	39370.1	3280.84	1093.61	1000*	1*	0.621371
1.60934×10^6*	160934	63360*	5280*	1760*	1609.34	1.60934	1*

1 thou = *0.0254 mm
1 Å (ångström) = 10^{-10}m
1 UK nautical mile = 6080 ft = 1853.2 m
1 international nautical mile = 6076.1 ft = *1852 m
1 μm (micron) = 10^{-6}m = 39.37×10^{-6} in
Note: starred numbers are exact conversions

Area

mm²	cm²	in²	ft²	yd²	m²	acre	(hectare) ha	km²	mile²
1	0.01	1.550×10^{-3}	1.076×10^{-5}	1.196×10^{-6}	10^{-6}				
100	1	0.1550	1.076×10^{-3}	1.196×10^{-4}	10^{-4}				
645.16	6.4516	1	6.944×10^{-3}	7.716×10^{-4}	6.452×10^{-4}				
92903	929	144	1	0.1111	0.09290	2.30×10^{-5}	9.29×10^{-6}	9.29×10^{-8}	3.587×10^{-8}
836127	8361	1296	9	1	0.8361	2.066×10^{-4}	8.361×10^{-5}	8.361×10^{-7}	3.228×10^{-7}
10^{6}	10000	1550	10.764	1.196	1	2.471×10^{-4}	10^{-4}	10^{-6}	3.861×10^{-7}
			43560	4840	4047	1	0.4047	4.047×10^{-3}	1.562×10^{-3}
			107639	11960	10000	2.471	1	0.01	3.861×10^{-3}
			1.0764×10^{7}	1.196×10^{6}	10^{6}	247.1	100	1	0.3861
			2.7878×10^{7}	3.0976×10^{6}	2.590×10^{6}	640	259.0	2.590	1

1 are = 100 m²

Volume

mm³	ml	in³	l*	US gal	UK gal	ft³	yd³	m³
1	1.0000×10^{-3}	6.1024×10^{-5}	1.0000×10^{-6}	2.642×10^{-7}	2.200×10^{-7}	3.531×10^{-8}	1.308×10^{-9}	10^{-9}
1000.0	1	0.061026	10^{-3}	2.642×10^{-4}	2.200×10^{-4}	3.532×10^{-5}	1.308×10^{-6}	1.0000×10^{-6}
16387	16.39	1	0.01639	4.329×10^{-3}	3.605×10^{-3}	5.787×10^{-4}	2.143×10^{-5}	1.639×10^{-5}
1.0000×10^{6}	1000	61.026	1	0.2642	0.2200	0.03532	1.308×10^{-3}	1.0000×10^{-3}
3.785×10^{6}	3785	231.0	3.785	1	0.8327	0.1337	4.951×10^{-3}	3.785×10^{-3}
4.546×10^{6}	4546	277.4	4.546	1:201	1	0.1605	5.946×10^{-3}	4.546×10^{-3}
2.832×10^{7}	2.832×10^{4}	1728	28.32	7.4805	6.229	1	0.03704	0.02832
7.6455×10^{8}	7.6453×10^{5}	46656	764.53	202.0	168.2	27	1	0.76456
10^{9}	1.0000×10^{6}	61024	1000.0	264.2	220.0	35.31	1.308	1

* 1 l = 1.000028 dm³ and 1 ml = 1.000028 cm³ according to the 1901
   definition of the liter
1 US barrel = 42 US gal = 34.97 UK gal
1 fluid oz = 28.41 ml
1 UK pint = 568.2 ml
1 liter = 1.760 UK pints

Volume Rate of Flow (Volume/Time)

litres/h	ml/s	m³/d	l/min	m³/h	ft³/min	l/s	ft³/s	m³/s
1	0.2778	0.024	0.01667	1×10^{-3}	5.886×10^{-4}	2.778×10^{-4}	9.810×10^{-6}	2.778×10^{-7}
3.6	1	0.08640	0.0600	3.6×10^{-3}	2.119×10^{-3}	1×10^{-3}	3.532×10^{-5}	1×10^{-6}
4.546	1.263	0.1091	0.07577	4.546×10^{-3}	2.676×10^{-3}	1.263×10^{-3}	4.460×10^{-5}	1.263×10^{-6}
41.67	11.57	1	0.6944	0.04167	0.02452	0.01157	4.087×10^{-4}	1.157×10^{-5}
60	16.67	1.44	1	0.0600	0.03531	0.01667	5.886×10^{-4}	1.667×10^{-5}
272.8	75.77	6.547	4.546	0.2728	0.1605	0.07577	2.676×10^{-3}	7.577×10^{-5}
1000	277.8	24	16.67	1	0.5886	0.2778	9.810×10^{-3}	2.778×10^{-4}
1699	471.9	40.78	28.31	1.699	1	0.4719	0.01667	4.719×10^{-4}
3600	1000	86.40	60	3.6	2.119	1	0.03531	1×10^{-3}
1.019×10^{5}	2.832×10^{4}	2446	1699	101.9	60	28.32	1	0.02832
1.894×10^{5}	5.261×10^{4}	4546	3157	189.4	111.5	52.61	1.858	0.05261
3.6×10^{6}	1×10^{6}	8.64×10^{4}	6×10^{4}	3600	2119	1000	35.31	1

Mass

g	oz	lb	kg	cwt	US ton (short ton)	t (tonne)	UK ton
1	0.035274	2.2046×10^{-3}	10^{-3}				
28.3495	1	0.0625	0.028350				
453.592	16	1	0.453592	8.9286×10^{-3}	5.00×10^{-4}	4.5359×10^{-4}	4.4643×10^{-4}
10^3	35.2740	2.20462	1	0.019684	1.1023×10^{-3}	10^{-3}	9.8421×10^{-4}
50802.3	1792	112	50.8023	1	0.056	0.05080	0.05
907185	32000	2000	907.185	17.8571	1	0.907185	0.892857
10^6	35273.9	2204.62	1000	19.6841	1.10231	1	0.984207
1.01605×10^6	35840	2240	1016.05	20	1.12	1.01605	1

1 quintal = 100 kg

Mass Rate of Flow (Mass/Time)

lb/h	kg/h	g/s	lb/min	lb/s	kg/s
0.2516	0.1142	0.03171	4.194×10^{-3}	6.990×10^{-5}	3.171×10^{-5}
0.2557	0.1160	0.03222	4.262×10^{-3}	7.103×10^{-5}	3.221×10^{-5}
1	0.4536	0.1260	0.01667	2.778×10^{-4}	1.260×10^{-4}
2.205	1	0.2778	0.03674	6.124×10^{-4}	2.778×10^{-4}
7.937	3.6	1	0.1323	2.205×10^{-3}	1×10^{-3}
60	27.216	7.560	1	1.667×10^{-2}	7.56×10^{-3}
91.86	41.67	11.57	1.531	0.02551	0.01157
93.33	42.34	11.76	1.556	0.02593	0.01176
2205	1000	277.8	36.74	0.6124	0.2778
2240	1016	282.2	37.33	0.6222	0.2822
3600	1633	453.6	60	1	0.4536
7937	3600	1000	132.3	2.205	1

Density (Mass/Volume)

kg/m³	lb/ft³	lb/in³	g/cm³
1	0.062428	3.6046×10^{5}	10^{-3}
16.0185	1	5.7870×10^{-4}	0.0160185
99.776	6.22884	3.6046×10^{-3}	0.099776
1000	62.4280	0.036127	1
1328.94	82.9630	0.048011	1.32894
27679.9	1728	1	27.6799

*1 g/cm³ = 1 kg/dm³ = 1 t/m³ = 1.000028 g/ml or 1.00028 kg/liter
(based on the 1901 definition of the liter)

Velocity

mm/s	ft/min	cm/s	km/h	ft/s	mile/h	m/s	km/s
1	0.19685	0.1	3.6×10^{-3}	3.281×10^{-3}	2.237×10^{-3}	10^{-3}	10^{-6}
5.08	1	0.508	0.018288	0.016667	0.01136	5.08×10^{-3}	5.08×10^{-6}
10	1.9685	1	0.036	0.032808	0.022369	0.01	10^{-5}
277.778	54.6806	27.7778	1	0.911344	0.621371	0.277778	2.778×10^{-4}
304.8	60	30.48	1.09728	1	0.681818	0.3048	3.048×10^{-4}
447.04	88	44.704	1.609344	1.46667	1	0.44704	4.470×10^{-4}
1000	196.850	100	3.6	3.28084	2.23694	1	10^{-3}
10^{6}	196850	100000	3600	3280.84	2236.94	10^{3}	1

1 UK knot = 1.853 km/h
1 international knot (Kn) = *1.852 km/h
Note: starred numbers are exact conversions

Second Moment of Area

mm⁴	cm⁴	in⁴	ft⁴	m⁴
1	10^{-4}	2.4025×10^{-6}	1.159×10^{-10}	10^{-12}
10000	1	0.024025	1.159×10^{-6}	10^{-8}
416231	41.623	1	4.8225×10^{-5}	4.1623×10^{-7}
8.631×10^{9}	863097	20736	1	8.6310×10^{-3}
10^{12}	10^{8}	2.4025×10^{6}	115.86	1

Force

pdl	N	lbf	kg f *	kN
1	0.1383	0.0311	0.0141	1.383×10^{-4}
7.233	1	0.2248	0.1020	10^{-3}
32.174	4.448	1	0.4536	4.448×10^{-3}
70.93	9.807	2.2046	1	9.807×10^{-3}
7233	1000	224.8	102.0	1
72070	9964	2240	1016	9.964

* The kgf is sometimes known as the kilopond (kp)

Moment of Force (Torque)

pdl ft	lbf in	N m	lbf ft	kgf m
1	0.3730	0.04214	0.03108	4.297×10^{-3}
2.681	1	0.1130	0.08333	0.01152
23.73	8.851	1	0.7376	0.1020
32.17	12	1.356	1	0.1383
232.7	86.80	9.807	7.233	1
6006	2240	253.1	186.7	25.81
72070	26880	3037	2240	309.7

One N m = 10^{7} dyn cm

Stress

dyn/cm²	N/m²	pdl/ft²	lbf/ft²	kN/m²	lbf/in²	kgf/cm²	MN/m²	kgf/mm²	h bar
1	0.100	0.06720	2.089×10^{3}	1×10^{-4}	1.450×10^{-5}	1.020×10^{-6}	1×10^{-7}	1.020×10^{-8}	1×10^{-8}
10	1	0.6720	0.02089	1×10^{-3}	1.450×10^{-4}	1.020×10^{-5}	1×10^{-6}	1.020×10^{-7}	1×10^{-7}
14.88	1.488	1	0.03108	1.488×10^{-3}	2.158×10^{-4}	1.518×10^{-5}	1.488×10^{-6}	1.518×10^{-7}	1.488×10^{-7}
478.8	47.88	32.17	1	0.04788	6.944×10^{-3}	4.882×10^{-4}	4.788×10^{-5}	4.882×10^{-6}	4.788×10^{-6}
1×10^{4}	1000	672.0	20.89	1	0.1450	0.01020	1×10^{-3}	1.020×10^{-4}	1×10^{-4}
6.895×10^{4}	6895	4533	144	6.895	1	0.07031	6.895×10^{-3}	7.031×10^{-4}	6.895×10^{-4}
9.807×10^{5}	9.807×10^{4}	6.590×10^{4}	2048	98.07	14.22	1	0.09807	0.01000	9.807×10^{-3}
1×10^{7}	1×10^{6}	6.720×10^{5}	2.089×10^{4}	1000	145.0	10.20	1	0.1020	0.1000
9.807×10^{7}	9.807×10^{6}	6.590×10^{6}	2.048×10^{5}	9807	1422	100	9.807	1	0.9807
1×10^{8}	1×10^{7}	6.720×10^{6}	2.089×10^{5}	10000	1450	102.0	10	1.020	1
1.544×10^{8}	1.544×10^{7}	1.038×10^{7}	3.226×10^{5}	1.544×10^{4}	2240	157.5	15.44	1.575	1.544

* 1 MN/m² = 1 N/mm²

Pressure

dyn/cm²	N/m²	lbf/ft²	m bar	mm Hg	in H₂O	kN/m²	in Hg	lbf/in²	kgf/cm²*	bar	atm
1	0.1000	2.089×10^{-3}	1×10^{-3}	7.501×10^{-4}	4.015×10^{-4}	1×10^{-4}	2.953×10^{-5}	1.450×10^{-5}	1.020×10^{-6}	1×10^{-6}	9.869×10^{-7}
10	1	0.02089	0.0100	7.501×10^{-3}	4.015×10^{-3}	1×10^{-3}	2.953×10^{-4}	1.450×10^{-4}	1.020×10^{-5}	1×10^{-5}	9.869×10^{-6}
478.8	47.88	1	0.4788	0.3591	0.1922	0.04788	0.01414	6.944×10^{-3}	4.882×10^{-4}	4.788×10^{-4}	4.726×10^{-4}
1000	100	2.089	1	0.7501	0.4015	0.1000	0.02953	0.01450	1.020×10^{-3}	1×10^{-3}	9.869×10^{-4}
1333	133.3	2.785	1.333	1	0.5352	0.1333	0.03937	0.01934	1.360×10^{-3}	1.333×10^{-3}	1.316×10^{-3}
2491	249.1	5.202	2.491	1.868	1	0.2491	0.07356	0.03613	2.540×10^{-3}	2.491×10^{-3}	2.458×10^{-3}
1×10^{4}	1000	20.89	10	7.501	4.015	1	0.2953	0.1450	0.01020	0.0100	9.869×10^{-3}
3.386×10^{4}	3386	70.73	33.86	25.40	13.60	3.386	1	0.4912	0.03453	0.03386	0.03342
6.895×10^{4}	6895	144	68.95	51.71	27.68	6.895	2.036	1	0.07031	0.06895	0.06805
9.807×10^{5}	9.807×10^{4}	2048	980.7	735.6	393.7	98.07	28.96	14.22	1	0.9807	0.9678
1×10^{6}	1×10^{5}	2089	1000	750.1	401.5	100	29.53	14.50	1.020	1	0.9869
1.013×10^{6}	1.013×10^{5}	2116	1013	760.0	406.8	101.3	29.92	14.70	1.033	1.013	1

* 1 kgf/cm² = 1 kp/cm² = 1 technical atmosphere
1 torr = 1 mm Hg (to within 1 part in 7 million)
1 N/m² is sometimes called a pascal

Energy, Work, Heat

J	ft lbf	*cal	kgf m	kJ	Btu	Chu	*kcal	MJ	hp h	kW h	therm
1	0.7376	0.2388	0.1020	10^{-3}	9.478×10^{-4}	5.266×10^{-4}	2.388×10^{-4}	10^{-6}	3.725×10^{-7}	2.778×10^{-7}	9.478×10^{-9}
1.3558	1	0.3238	0.1383	1.356×10^{-3}	1.285×10^{-3}	7.139×10^{-4}	3.238×10^{-4}	1.356×10^{-6}	5.051×10^{-7}	3.766×10^{-7}	1.285×10^{-8}
4.1868	3.0880	1	0.4270	4.187×10^{-3}	3.968×10^{-3}	2.205×10^{-3}	10^{-3}	4.187×10^{-6}	1.560×10^{-6}	1.163×10^{-6}	3.968×10^{-8}
9.8066	7.2330	2.3420	1	9.807×10^{-3}	9.294×10^{-3}	5.163×10^{-3}	2.342×10^{-3}	9.807×10^{-6}	3.653×10^{-6}	2.724×10^{-6}	9.294×10^{-8}
1000	737.56	238.85	101.97	1	0.9478	0.5266	0.2388	10^{-3}	3.725×10^{-4}	2.778×10^{-4}	9.478×10^{-6}
1055.1	778.17	252.00	107.59	1.0551	1	0.5556	0.2520	1.055×10^{-3}	3.930×10^{-4}	2.931×10^{-4}	10^{-5}
1899.1	1400.7	453.59	193.71	1.8991	1.800	1	0.4536	1.899×10^{-3}	7.074×10^{-4}	5.275×10^{-4}	1.800×10^{-5}
4186.8	3088.0	1000	427.04	4.1868	3.9683	2.2046	1	4.187×10^{-3}	1.560×10^{-3}	1.163×10^{-3}	3.968×10^{-5}
10^{6}	737 562	238 846	101 972	1000	947.82	526.56	238.85	1	0.3725	0.2778	9.478×10^{-3}
2.6845×10^{6}	1.9800×10^{6}	641 186	273 745	2684.5	2544.4	1413.6	641.19	2.6845	1	0.7457	0.03544
3.6000×10^{6}	2.65522×10^{6}	859 845	367 098	3600	3412.1	1895.6	859.84	3.600	1.3410	1	0.03412
1.0551×10^{8}	7.7817×10^{7}	2.5200×10^{7}	1.0759×10^{7}	1.0551×10^{5}	100 000	55 556	25200	105.51	39.301	29.307	1

* cal is the International Table calorie
1 hp h = 1.014 hp h (metric) = 745.7 Wh = 2.685 MJ
1 thermie = 1.163 kWh = 4.186 MJ = 999.7 kcal
1 ft pdl = 0.04214 J
1 erg = 10^{-7} J

Power, Heat Flow Rate

Btu/h	Chu/h	W	kcal/h	ft lbf/s	kgf m/s	metric hp	hp	kW	tcal/h	MW
1	0.5556	0.2931	0.2520	0.2162	0.02988	3.985×10^{-4}	3.930×10^{-4}	2.931×10^{-4}	2.520×10^{-4}	2.93×10^{-7}
1.800	1	0.5275	0.4536	0.3892	0.05379	7.172×10^{-4}	7.074×10^{-4}	5.275×10^{-4}	4.536×10^{-4}	5.28×10^{-7}
3.4121	1.8956	1	0.8598	0.7376	0.1020	1.360×10^{-3}	1.341×10^{-3}	10^{-3}	8.598×10^{-4}	10^{-6}
3.9683	2.2046	1.163	1	0.8578	0.1186	1.581×10^{-3}	1.560×10^{-3}	1.163×10^{-3}	10^{-3}	1.16×10^{-6}
4.626	2.5701	1.3558	1.1658	1	0.1383	1.843×10^{-3}	1.818×10^{-3}	1.356×10^{-3}	1.166×10^{-3}	1.36×10^{-6}
33.46	18.59	9.807	8.432	7.233	1	0.01333	0.01315	9.807×10^{-3}	8.432×10^{-3}	9.81×10^{-6}
2510	1394	735.50	632.4	542.5	75	1	0.9863	0.7355	0.6324	7.355×10^{-4}
2544	1414	745.70	641.19	550	76.040	1.0139	1	0.7457	0.6412	7.457×10^{-4}
3412.1	1896	1000	859.8	737.6	102.0	1.360	1.3410	1	0.8598	10^{-3}
3968.3	2204.6	1163	1000	857.8	118.6	1.581	1.5596	1.163	1	1.163×10^{-3}
3.4121×10^6	1.896×10^6	10^6	8.598×10^5	7.376×10^5	1.0197×10^5	1360	1341	1000	859.8	1

1 W = 1 J/S
1 cal/s = 3.6 kcal/h
1 ton of refrigeration = 3517 W = 12000 Btu/h
1 erg/s = 10^{-7} W

Dynamic Viscosity

μN s/m²	kg/mh	lb/ft h	cP	P	Ns/m²	pdl s/ft²	kgf s/m²	lbf s/ft²	lbf h/ft²
1	3.6×10^{-3}	2.419×10^{-3}	10^{-3}	10^{-5}	10^{-6}	6.720×10^{-7}	1.020×10^{-7}	2.09×10^{-8}	5.80×10^{-12}
277.8	1	0.6720	0.2778	2.778×10^{-3}	2.778×10^{-4}	1.866×10^{-4}	2.832×10^{-5}	5.80×10^{-6}	1.61×10^{-9}
413.4	1.488	1	0.4134	4.134×10^{-3}	4.134×10^{-4}	2.776×10^{-4}	4.215×10^{-5}	8.63×10^{-6}	2.40×10^{-9}
1000	3.6	2.419	1	0.010	10^{-3}	6.720×10^{-4}	1.020×10^{-4}	2.09×10^{-5}	5.80×10^{-9}
10^5	360	241.9	100	1	0.100	0.0672	0.01020	2.089×10^{-3}	5.80×10^{-7}
10^6	3600	2419	1000	10	1	0.6720	0.1020	0.02089	5.80×10^{-6}
1.488×10^6	5358	3600	1488	14.88	1.488	1	0.15175	0.03108	8.63×10^{-6}
9.807×10^6	3.530×10^4	2.372×10^4	9807	98.07	9.807	6.590	1	0.2048	5.69×10^{-5}
4.788×10^7	1.724×10^5	1.158×10^5	47880	478.8	47.88	32.17	4.882	1	2.778×10^{-4}
1.724×10^{11}	6.205×10^8	4.170×10^8	1.724×10^8	1.724×10^6	1.724×10^5	1.153×10^5	1.758×10^4	3600	1

1 cp = 1 mN s/m² = 1 g/m s
1 P = 1 g/cm s = 1 dyn s/cm²
1 N s/m² = 1 kg/m s
1 pdl s/ft² = 1 lb/ft s
1 lbf s/ft² = 1 slug/ft s

Kinematic Viscosity

in²/h	cSt (mm²/s)	ft²/h	St (cm²/s)	m²/h	in²/s	ft²/s	m²/s
1	0.1792	6.944×10^{-3}	1.792×10^{-3}	6.452×10^{-4}	2.778×10^{-4}	1.93×10^{-6}	1.79×10^{-7}
5.5800	1	0.03875	0.010	3.60×10^{-3}	1.550×10^{-3}	1.076×10^{-5}	10^{-6}
144	25.81	1	0.2581	0.0929	0.04	2.778×10^{-4}	0.258×10^{-4}
558.0	100	3.8750	1	0.36	0.1550	1.076×10^{-3}	10^{-4}
1550	277.8	10.76	2.778	1	0.4306	2.990×10^{-3}	2.778×10^{-4}
3600	645.2	25	6.452	2.323	1	6.944×10^{-3}	6.452×10^{-4}
518400	92903	3600	929.0	334.5	144	1	0.0929
5.580×10^{6}	10^{6}	3.875×10^{4}	10000	3600	1550	10.76	1

Density of Heat Flow Rate (Heat/Area × Time)

W/m²	kcal/m²h	Btu/ft²h	Chu/ft²h	kcal/ft²h	kW/m²
1	0.8598	0.3170	0.1761	0.07988	10^{-3}
1.163	1	0.3687	0.2049	0.09290	1.163×10^{-3}
3.155	2.712	1	0.5556	0.2520	3.155×10^{-3}
5.678	4.882	1.800	1	0.4536	5.678×10^{-3}
12.52	10.76	3.968	2.205	1	0.01252
1000	859.8	317.0	176.1	79.88	1

Heat-Transfer Coefficient
(Thermal Conductance; Heat/Area × Time × Degree Temperature)

W/m² °C	kcal/m²h°C	Btu/ft²h °F	kcal/ft²h °C	kW/m² °C	Btu/ft²s °F	cal/cm²s°C
1	0.8598	0.1761	0.07988	10^{-3}	4.892×10^{-5}	2.388×10^{-5}
1.163	1	0.2048	0.09290	1.163×10^{-3}	5.689×10^{-5}	2.778×10^{-5}
5.678	4.882	1	0.4536	5.678×10^{-3}	2.778×10^{-4}	1.356×10^{-4}
12.52	10.76	2.205	1	0.01252	6.124×10^{-4}	2.990×10^{-4}
1000	859.8	176.1	79.88	1	0.04892	0.02388
20442	17577	3600	1633	20.44	1	0.4882
41868	36000	7373	3344	41.87	2.048	1

1 Btu/ft² h°F = Chu/ft²h ° C
1 W/m²°C = 10^{-4} W/cm²° C

Thermal Conductivity
(Heat × Length/Area × Time × Degree Temperature)

Btu in/ft²h°F	kcal in/ft²h C	W/m°C	kcal/m h°C	Btu/ft h°F	cal/cm s°C
1	0.4536	0.1442	0.1240	0.0833	3.445×10^{-4}
2.2046	1	0.3180	0.2734	0.1837	7.594×10^{-4}
6.933	3.146	1	0.8598	0.5778	2.388×10^{-3}
8.064	3.658	1.163	1	0.6720	2.778×10^{-3}
12	5.443	1.731	1.488	1	4.134×10^{-3}
2903	1317	418.7	360	241.9	1

1 Btu In/ft²h °F = 1 Chu in/ft²h °C
1 Btu/ft h °F = 1 Btu ft/ft²h °F = 1 Chu/ft h °C
1 W/m °C = 10^{-2} W/cm °C = 1 kW mm/m² °C

Specific Heat Capacity (Heat/Mass × Degree Temperature)

ft lbf/lb°F	kgf m/kg°C	kJ/kg°C	*Btu/lb°F	kcal/kg°C
1	0.5486	5.380×10^{-3}	1.285×10^{-3}	1.285×10^{-3}
1.823	1	9.807×10^{-3}	2.342×10^{-3}	2.342×10^{-3}
185.9	101.97	1	0.2388	0.2388
778.2	426.9	4.1868	1	1
778.2	426.9	4.1868	1	1

* 1 Btu/lb °F = 1 Chu/lb °C

Specific Energy
(Heat/Mass; e.g., Calorific Value, Mass Basis, Specific Latent Heat)

ft lbf / lb	kgf m/kg	*kJ/kg	Btu/lb	kcal / kg	MJ / kg
1	0.3048	2.989×10^{-3}	1.285×10^{-3}	7.139×10^{-4}	2.989×10^{-6}
3.281	1	9.807×10^{-3}	4.216×10^{-3}	2.342×10^{-3}	9.807×10^{-6}
334.55	101.97	1	0.4299	0.2388	10^{-3}
778.2	237.19	2.326	1	0.5556	2.326×10^{-3}
1400.7	426.9	4.187	1.8	1	4.187×10^{-3}
334553	101972	1000	429.9	238.8	1

* 1 J/g = 1 kJ/kg   1 kcal/kg = 1 Chu/lb

Calorific Value, Volume Basis (Heat/Volume)

J/m³	kJ/m³	kcal /m³	Btu/ft³	Chu/ft³	*MJ / m³
1	1×10^{-3}	2.388×10^{-4}	2.684×10^{-5}	1.491×10^{-5}	1×10^{-6}
1000	1	0.2388	0.02684	0.01491	1×10^{-3}
4.187×10^{3}	4.187	1	0.1124	0.06243	4.187×10^{-3}
3.726×10^{4}	37.26	8.899	1	0.5556	0.03726
6.707×10^{4}	67.07	16.02	1.800	1	0.06707
1×10^{6}	1000	238.8	26.84	14.91	1

1 therm (10^5 Btu) / UK gal = 2320 8 MJ/m³
1 thermie/liter = 4185 MJ/m³
* MJ/m³ = J/cm³

References

1. Lok, H.H. "Untersuchungen an Dichtungun für Apparateflansche," Diss. Tech. High School, Delft, 1960.

2. "Klinger takes the Guesswork out of Gaskets," Klinger International Export GmbH, P.O. Box 24, A-1125 Vienna, Austria, 1969.

3. Boon, E.F. "Fundamentals of Flange and Shaft Seals," *Dechema-Monographien*, pp. 372-392 (1952).

4. Warrick, R.V. "The Good Business Role of Valve Standards," *Pipe-line Engineer* (July 1963).

5. Hill, R., E.H. Lee and S.J. Tupper. "Method of Numerical Analysis of Plastic Flow in Plain Strain and Its Application to the Compression of a Ductile Material between Rough Plates," *J. Appl. Mech.*, p. 49 (June 1951).

6. Diederich, H. and V. Schwarz. *The Optimal Application of Butterfly Valves*, VAG-Armaturen GmbH (translation from *J. Schiff und Hafen*, No. 22, pp. 740-741. Seehafen-Verlag Erik Blumenfeld, Hamburg, (August 1970).

7. Krägeloh, E. "Die Wesentlichsten Prüfmethoden für It-Dichtungen," *J. Gummi und Asbest — Plastische Massen* (November 1955).

8. Morrison, J.B. "O-Rings and Interference Seals for Static Applications," *Machine Design*, pp. 91-94 (February 7, 1957).

9. Gillespie, L.H., D.O. Saxton and P.M. Chapmen. "New Design Data for FEP, TFE," *Machine Design* (January 21, 1960).

10. Turnbull, D.E. "The Sealing Action of a Conventional Stuffing Box," *Brit. Hydr. Res. Assoc.,* RR 592 (July 1958).

11. Denny, D.F. and D.E. Turnbull. "Sealing Characteristic of Stuffing Box Seals for Rotating Shafts," *Proc. Instn. Mechn. Engrs.,* London, Vol. 174, No. 6 (1960).

12. Reynolds, H.J. Jr. "Mechanism of Corrosion of Stainless Steel Valve Stems by Packing — Methods of Prevention," Johns-Manville Prod. Corp. (Preprint No. 01-64 for issue by API Division Refining).

13. Darvas, L.A. "Cavitation in Closed Conduit Flow Control Systems," *Civil Engg. Trans., Instn. Engrs. Aust.,* Vol. CE12, No. 2, pp. 213-219 (October 1970).

14. O'Brien, T. "Needle Valves with Abrupt Enlargements for the Control of a High Head Pipeline," *J. Instn. Engrs. Aust.,* Vol. 38, Nos. 10-11, pp. 265-274 (October/November 1966).

15. Krost, H., K. Kuckelhaus and J. Sulliga. "Combined Start-Up, Pressure Reduction, and Safety Valves for Big Power Plant Sets," *Combustion,* pp. 32-41 (September 1971).

16. Streeter, V.L. and E.B. Wylie. "Fluid Transients," NY: McGraw-Hill Book Company.

17. Selle, H. "Die Zündgefahren bei Verwendung verbrennlicher Schmier — und Dichtungsmittel für Sauerstoff-Hochdruck-armaturen," *J. Die Berufsgemeinschaft,* (October 1951).

18. Rasmussen, L.M. "Corrosion by Valve Packing," *Corrosion,* Vol. 11, No. 4, pp. 25-40 (1955).

19. Uhlig, Herbert H. "Corrosion and Corrosion Control: An Introduction to Corrosion Science and Engineering," NY: John Wiley and Sons, Inc.

20. Evans, U.R. "The Corrosion and Oxidation of Metals," NY: St. Martins Press, Inc.

21. "Corrosion and Wear Handbook for Water Cooled Reactors," United States Atomic Energy Commission, Washington, edited by D.J. Paul (March 1957).

22. La Que, F.L. and H.R. Copson. "Corrosion Resistance of Metals and Alloys," Reinhold Publishing Corp., Chapman Hall Ltd.

23. "Corrosion Data Survey," NACE (1973).

24. "A Guide to Corrosion Resistance," Climax Molybdenum Co.

25. Ingard, U. "Attenuation and Regeneration of Sound in Ducts and Jet Diffusers," *J.Ac.Soc.Am.*, Vol. 31, No. 9, pp. 1202-1212 (1959).

26. Weiner, R.S. "Basic Criteria and Definitions for Zero Fluid Leakage," Technical Report No. 32-926, *Jet Propulsion Laboratory Cal. Instn. of Techn.*, Pasadena, CA (December 1966).

27. "Spiral Wound Gaskets Design Criteria," Flexitallic Bulletin No. 171, Flexitallic Gasket Co., Inc. (1971).

28. Pool, E.B. "Minimization of Surge Pressure from Check Valves for Nuclear Loops," ASME Publication 62-WA-220.

29. Pool, E.B., A.J. Porwit and J.L. Carlton. "Prediction of Surge Pressure from Check Valves for Nuclear Loops," ASME Publication 62-WA-219.

30. "Pressure Losses in Valves," *Engineering Sciences Data Item No. 69022*, Engineering Sciences Data Unit, London.

31. Thompson, L. and O.E. Buxton Jr. "Maximum Isentropic Flow of Dry Saturated Steam through Pressure Relief Valves," *J. of Pressure Vessel Technology*, Vol. 101, pp. 113-117 (May 1979).

32. Lapple, C.E. "Isothermal and Adiabatic Flow of Compressed Fluids," *Trans. AIChE*, Vol. 39, pp. 385-432 (1943).

33. Duffey, D.W. and E.A. Bake. "A Hermetically Sealed Valve for Nuclear Power Plant Service," V-Rep. 74-3, Flow Control Division Rockwell International.

34. Kreuz, A. "Einfluss von Gegendrücken auf Verschiedene Bauarten von Sicherheitsventilen," *Armaturen International 68*, pp. 17-19, Exportzeitschriften-Union Düsseldorf.

35. Wells, F.E. "A Survey of Leak Detection for Aerospace Hardware," paper presented at the National Fall Conference for the American Society of Nondestructive Testing, Detroit, October 1968.

36. "Recommended Practice for the Design and Installation of Pressure-Relieving Systems in Refineries — Design," API RP 520 Part I (1976).

37. "Recommended Practice for the Design and Installation of Pressure-Relieving Systems in Refineries — Installation," API RP 520 Part II (1963).

38. Randall, R.B. Brüel & Kjaer, Copenhagen, private communications.

39. Conison, J. "How to Design a Pressure Relief System," *Chem. Engg.* (July 25, 1960).

40. Holmberg, E.G. "Valve Design: Special for Corrosives," *Chem. Engg.* (June 13, 1960).

41. Holmberg, E.G. "Valves for Severe Corrosive Service," *Chem. Engg.* (June 27, 1960).

42. Champagne, R.P. "Study Sheds New Light on Whether Increased Packing Height Seals a Nuclear Valve Better," *Power* (May 1976).

43. Velan, A.K. "New Test Work Sheds Light on Criteria for Nuclear Valve Stem Seals and Packing," *Power* (December 1973).

44. Awtrey, P.H. "Pressure-Temperature Ratings of Steel Valves," *Heating/Piping/Air Conditioning,* pp. 109-114 (May 1978).

45. Allen, E.E. "Control Valve Noise," ISA Handbook of Control Valves (1976).

46. Baumann, H.D. "Universal Valve Noise Prediction Method," ISA Handbook of Control Valves (1976).

47. Arant, J.B. "Coping with Control Valve Noise," ISA Handbook of Control Valves (1976).

48. Scull, W.L. "Control Valve Noise Rating: Prediction versus Reality," ISA Handbook of Control Valves (1976).

49. Sparks, C.R. and D.E. Lindgreen. "Design and Performance of High-Pressure Blowoff Silencers," *J. of Engg. for Industry* (May 1971).

50. Chesler, S. and B.W. Jesser. "Some Aspects of Design and Economic Problems Involved in Safe Disposal of Inflammable Vapors from Safety Relief Valves," *Trans. ASME,* Vol. 74, pp. 229-246 (1952).

51. Bull, M.K. and D.C. Rennison. "Acoustic Radiation from Pipes with Internal Turbulent Gas Flows," proceedings from the Noise, Shock & Vibration Conference, Monash University, Melbourne, pp. 393-405 (1974).

52. Bull, M.K. and M.P. Norton. "Effects of Internal Flow Disturbances on Acoustic Radiation from Pipes," proceedings from the Vibration and Noise Control Engineering Conference, Instn. Engrs. Aust., Sydney, pp. 61-65 (1976).

53. German Guidelines for Unfired Pressure Vessels: Sicherheitsventile, AD-Merkblatt A2, Beuth-Vertrieb GmbH, Berlin W30 (1980).

54. German Guidelines for Steam Boilers: Sicherheitseinrichtungen gegen Drucküberschreitung, SR-Sicherheitsventile, Beuth-Vertrieb GmbH, Berlin W30 (1972).

55. VDI-Wasserdampftafel, 7. Auflage, VDI-Verlag, Düsseldorf (1968).

56. "Nonmandatory Rules for the Design of Safety Relief Valve Installations," ANSI B31.1-1977 Power Piping, Appendix II.

57. Cooper, Walter. "A Fresh Look at Spring Loaded Packing," *Chemical Engineering,* pp. 278-284 (November 6, 1967).

58. Ball, J.W. "Cavitation Characteristics of Gate Valves and Globe Valves Used as Flow Regulators Under Heads up to about 125 Feet," *Trans. ASME,* Vol. 79, Paper No. 56-F-10, pp. 1275-1283 (August 1957).

59. Ball, J.W. "Sudden Enlargements in Pipelines," *Proc. ASCE Jour. Power Div.,* Vol. 88, No. P04, pp. 15-27 (December 1962).

60. Häfele, C.H. and A. Kreuz. "Rohrleitungsarmaturen," *Rohrleitungen, Theorie und Praxis,* S. Schwaigerer, ed., Springer-Verlag (1967).

61. Häfele, C.H. Sempell Armaturen, Korschenbroich (Germany), Private Communications.

62. Simin, O. "Water Hammer," *Proc. AWWA,* pp. 335-424 (1904), Summary of Work in Moscow by Prof. N. Joukowsky.

63. Hutchinson, J.W. "ISA Handbook of Control Valves," Instrument Society of America.

64. IEC Publication 534-1. "Industrial Control Valves, Part 1: General Considerations," International Electrotechnical Commission, Geneva, Switzerland.

65. ANSI/ISA 575.01. "Control Valve Sizing Equations."

66. Sallet, D.W. "On the Sizing of Pressure Relief Valves for Pressure Vessels which are used in the Transport of Liquified Gases," ASME Publication 78-WA/HT-39.

67. "Compression Packings Handbook," Fluid Sealing Association, 2017 Walnut Street, Philadelphia, Pa., 19103.

68. Kobori, T.S. Yokoyama, and H. Miyashiro. "Propagation Velocity of Pressure Wave in Pipe Line," *Hitachi Hyoron,* Vol. 37, No. 10, Oct. 1955.

69. Baha, D.Y. and J.G. Beese, "Optimum Pre-Doming of Metallic Safety-Rupture Diaphragms," Paper C 267/79, I Mech E Conference Publications, 1979-14.

70. Ens, H., Jr. "Rupture Discs as Pressure Relief Devices," Paper C268/79, I Mech E Conference Publications, 1979-14.

71. Harrison, S.F. Safety and Safety Relief Valve Requirements, The National Board of Boiler and Pressure Vessel Inspectors and the American Society of Mechanical Engineers, Paper 265/79, I Mech E Conference Publications, 1979-14.

72. Brodie, C.W. "Safety Discs, a Protective Device for Safety Valves," Paper C 266/79, I Mech E Conference Publications, 1979-14.

73. Brodie, G.W. "Development in the Design and Manufacture of Graphite Bursting Discs," Paper C 377/84, I Mech E Conference Publications, 1984-13.

74. Burns, Madgruber L., Jr. "The Practical Application of Flow Test Data for Safety Relief Valves and Rupture Discs used in Combination," Paper C 380/84, I Mech E Conference Publications, 1984-13.

75. "Roll-Over of Metal Reverse Bursting Discs: A Problem," Loss Prevention Bulletin Number 045, I Chem E, London.

76. "Rupture Disc Technical Manual," 2nd Edition, March 1, 1985, Continental Disc Corporation.

77. F.H. Bielesch. SIGRI GmbH, Meitingen, West Germany, Private Communications.

78. SVDB Band 2, Armaturen and Ausruestung-Sicherheitsventile, Vorschrift 602.

79. VdTUV-Specification Safety Valves 100/2, "Proposal for the Sizing of Safety Valves for Gases in the Liquid State."

Index